ETHERNET/IP™

The Everyman's Guide to The Industry's Leading Automation Protocol

JOHN S. RINALDI

With JAMIN WENDORF

JOHN S. RINALDI

Copyright © 2018 JOHN S. RINALDI

All rights reserved.

ISBN: 978-1726662567

THE EVERYMAN'S GUIDE TO ETHERNET/IP

Copyright Notice © 2018 Real Time Automation, Inc. All rights reserved. Printed in USA.

This document is copyrighted by Real Time Automation Inc. Any reproduction and/or distribution without prior written consent from Real Time Automation, Inc. is strictly prohibited.

Trademark Notices
Allen-Bradley, ControlLogix, FactoryTalk, PLC-5, Rockwell Automation, Rockwell Software, RSLinx, RSView, are registered trademarks of Rockwell Automation, Inc.

ControlNet is a registered trademark of ODVA Inc.
DeviceNet is a trademark of the ODVA Inc.
EtherNet/IP is a trademark of the ODVA Inc
Modbus TCP is a trademark of Modbus IDA
Ethernet is a registered trademark of Digital Equipment Corporation, Intel, and Xerox Corporation.
Microsoft, Windows, Windows ME, Windows NT, Windows 2000, Windows Server 2003, and Windows XP are either registered trademarks or trademarks of Microsoft Corporation in the United States and/or other countries.

All other trademarks are the property of their respective holders and are hereby acknowledged.

JOHN S. RINALDI

WHAT THE INDUSTRY EXPERTS ARE SAYING...

"John Rinaldi's 7th book is an illuminating illustration of what's going on in connected technology for the seasoned pro or the apprentice. His inner compass gives insight, foresight & hindsight on key technology tenets of Ethernet IP & OPC UA as foundational to deliver the connected enterprise in a monumental way. Enjoy the roller coaster read as John communicates lively & deeply on how Ethernet IP originated, who coordinated, who got eliminated, and how we landed at a place where it is interoperated with OPC UA & PackML."
Paul Redwood, Chair, OMAC; R&D Manager, Church & Dwight

"In the confusing industrial automation world where Ethernet is Ethernet and IP is IP, but Ethernet plus IP is not Ethernet/IP, this book is a clear and enjoyable guide to the protocol, its origins and current use, and its relationship to other standards. With his engaging style and unique wit, John Rinaldi demystifies Ethernet/IP in a way that is approachable and helpful to both beginners and veterans of industrial automation and networking."
John Traynor, Senior Vice President, FreePoint Technologies

"From the history of Ethernet/IP basics to where it lives today in the sphere of Industry 4.0, truly another in a series of publications from J. Rinaldi, that both educates and helps guide the layman and expert alike. Let both the OT and IT worlds take a mutually compelling journey, regarding Ethernet/IP's foundation and future."
Jeff Stanko, Technical Consultant and Industry 4.0 Specialist, McNaughton-McKay Electric

"This book is a great resource for understanding the technical aspects of ETHERNET/IP™ and its role in automation architectures that are driving greater manufacturing, and process industry productivity. John Rinaldi clearly communicates with an easy to understand and effective writing style, so readers benefit from his many years of practical experience."
Bill Lydon, Automation Industry Expert Consultant & Analyst, Editor Automation.com

"EtherNet/IP and the CIP protocol are important technologies for control, automation, and the Industrial Internet of Things. John has a talent for describing technology for decision-makers who may not be programmers. He did it for OPC UA and again here for EtherNet/IP."
Gary Mintchell, CEO and Founder The Manufacturing Connection

DEDICATION

*To the Automation Engineer,
the unsung hero of American Manufacturing*

JOHN S. RINALDI

TABLE OF CONTENTS

DEDICATION	5
TABLE OF FIGURES	7
TABLE OF TABLES	8
ACKNOWLEDGMENTS	9
FOREWORD	10
INTRODUCTION	13
WHAT IS INDUSTRIAL ETHERNET?	16
ETHERNET/IP HISTORY	22
CIP & EIP: TEN THINGS TO KNOW	23
WHAT IS CIP?	33
CIP OBJECT MODELING	34
CIP MESSAGING	52
CIP & ETHERNET/IP	54
TCP & ETHERNET/IP	63
THE ETHERNET/IP ADAPTER	69
THE ETHERNET/IP SCANNER	79
COMPLIANCE TESTING	102
ETHERNET/IP FOR END USERS	100
ETHERNET/IP FOR SW ENGINEERS	103
EDS FILES & CONFIGURATION	109
CIP EXTENSIONS	116
THE ODVA	121
ETHERNET/IP AND THE IOT	124
ETHERNET/IP VS PROFINET	130
ETHERNET/IP AND OPC UA	134
ETHERNET/IP AND DEVICENET	141
ETHERNET/IP AND RTA	145
ABOUT THE AUTHOR	147
OTHER BOOKS BY JOHN S RINALDI	149

THE EVERYMAN'S GUIDE TO ETHERNET/IP

TABLE OF FIGURES

Figure 1 – CIP Overview 30
Figure 2 – CIP Protocol Layers 31
Figure 3 – Modbus Address Space 36
Figure 4 – OPC UA Address Space 37
Figure 5 – Structure of Node with Objects 38
Figure 6 – CIP Object Class Example 39
Figure 7 – CIP Communication Objects 46
Figure 8 – Comm. Objects for Two Explicit Connections 47
Figure 9 – Implicit Messaging Communication Objects 48
Figure 10 – Logical Segment Example (Hex Values) 49
Figure 11 – Explicit Message Packet Contents 50
Figure 12 - Explicit Message Response Format 50
Figure 13 – CIP Implicit Input Message Packet 52
Figure 14 – CIP Implicit Output Message Format 52
Figure 15 – EtherNet/IP Object Structure 56
Figure 16 – CIP Explicit and Implicit Message Encapsulation 58
Figure 17 – CIP Encapsulation 59
Figure 18 – Some of The Many TCP/IP Protocol Layers 65
Figure 19 – How TCP/IP Protocols Layer On Each Other 66
Figure 20 – The EtherNet/IP Adapter Object Model 73
Figure 21 - Ethernet/IP Adapter Software Structure 77
Figure 22 - EtherNet/IP Scanner Object Model 82
Figure 23 - Connection Initiation for Connected Messages 86
Figure 24 - Connection Initiation for Unconnected Messages 88
Figure 25 – Ethernet/IP Scanner Software Structure 91
Figure 26 – ODVA Terms of Usage Agreement (1st Page) 96
Figure 27 – Ethernet/IP Technology Only Logo 99
Figure 28 – Ethernet/IP And ODVA Combined Logo 99
Figure 29 – Logo For Conformant Products 99
Figure 30 – ODVA Member Logo 99
Figure 31 – ODVA Conformance Logo 100
Figure 32 – OSI vs. Ethernet/IP Software Model 104
Figure 33 – EIP Object Model Physical Connections 106
Figure 34 - User Application Interface 108
Figure 35 - Sample EDS File (partial) 115
Figure 36 – ODVA Membership Logo 123
Figure 37 – Architecture to Move Data to the Cloud 127
Figure 38 – EIP and OPC UA Cloud Communication 128

Figure 39 – Proxy Based Communications to the Cloud 129
Figure 40 - EtherNet/IP vs. DeviceNet Protocol Layers 143

TABLE OF TABLES

Table 1 – CIP Device Types 31
Table 2 - Important CIP Terminology 32
Table 3 – Object, Instance, Attribute Identifiers 40
Table 4 – Some Common CIP Objects 41
Table 5 – The CIP Object Library (Partial) 42
Table 6 –Attributes of the Connection Object Class 49
Table 7 – Common CIP Service Codes 51
Table 8 – Service Code Identifier Ranges 51
Table 9 – EtherNet/IP Encapsulation Commands 60
Table 10 – Forward Open: Most Important Parameters 61
Table 11 – Five Important ODVA Branding Guidelines 98
Table 12 – EtherNet/IP vs. PROFINET Comparison 132

THE EVERYMAN'S GUIDE TO ETHERNET/IP

ACKNOWLEDGMENTS

This book would not be possible without the dedication, friendship, persistence, support and follow through of the entire staff at Real Time Automation. I am very grateful for all they do so that I can take on projects like this.

By reading and accepting this information you agree to all of the following: The information presented in this publication is for the general education of the reader. You understand that this is simply a set of opinions (and not advice). This is to be used for education and not considered as "professional" advice. You are responsible for any use of this information in this work and hold the author and all members and affiliates harmless in any claim or event. Both the author and the publisher disclaim any and all liability of any kind arising out of such use. The reader is expected to exercise sound professional judgment in using any of the information presented in any particular application.

JOHN S. RINALDI

FOREWORD

It's very exciting to get to write a forward for "The Everyman's Guide to EtherNet/IP", authored by the visionary and creative genius John Rinaldi.

John is a courageous free spirit and his astonishing reputation and admiration in industrial automation and Manufacturing is undisputable. He's not afraid to challenge the norm, and this book is another testimonial to his great storytelling ability and vision.

John tells the complete story with the facts and often stirs the world with respect to suppliers and consortium's intellectual boundaries. John always shares his excellent perspective on technology that often challenges the minds of some of the greatest leaders, of both the suppliers and consortium's (including myself!).

John's heart, soul and passion for technology, anticipating and solving the problems in manufacturing well in advance of the customer realizing their own needs is unquestionable.

This book is a further example of John's credibility in our industry, being able to be an effective leader and extremely effective communicator of the complete story of the EtherNet/IP technology. John provides a deep introspective providing examples and comparisons of the perceived competitive technology to EtherNet/IP.

I spent most of my career at Rockwell, but I can now say I have learned more about EtherNet/IP and all the earlier technologies of ControlNet and DeviceNet® from John's book than I learned during my whole career at Rockwell!

You have made the right decision to start reading this book and I guarantee you won't want to put it down before you completely understand the whole story of EtherNet/IP from both a business and technology perspective. You also will be indoctrinated with John's vision of the future.

This book is the icing on the cake with respect to the other books that John has authored in this series.

John can predict the future as he looks into his crystal ball and reads the minds of the customers before they've even thought of what they need to solve their manufacturing dilemmas approaching. John is able to look ahead and determine from a business value perspective how to leverage technology innovations that we haven't even dreamed of.

Now that you have this book in your hands, I highly encourage you to not put it down and make the time to read the book from cover to cover. Use this book as your reference manual for the future, whether you're a supplier or user system integrator, to truly understand the true story and the power of EtherNet/IP.

It's a sincere pleasure to know John personally and listen to his stories. He clearly is recognized as a leader and visionary in our industry. You now have the distinguished honor of being able to have some of his stories in print directly from his mouth and an opportunity to learn from the past, understand the current challenges of today and see what the future will bring through John's visionary pedigree and prophecies.

My vision is all about the business value proposition of applying the right technology from the myriad of technical innovations. Everyone is in a very challenging changing time as the suppliers and end-users are looking to maximize the use of technology as they look to raise revenue and decrease expenses. End-users are all looking at maximizing efficiency of operations. Manufacturers are looking for streamlined operations, inclusive of intuitive configuration, highly intuitive diagnostics, and easy ways to get data and information out of their field devices and into other disparate devices and applications. The importance of suppliers and end-users working together cannot be underestimated. I've seen quite a few trends where the end-users are demanding the suppliers truly embrace industry standards. One of the key standards and that John writes about in this book being EtherNet/IP is so important to manufacturing. There's also a recognition of working with other organizations and collaboration which John talks about concerning the OPC UA architecture, and how OPC UA has to get data out of the corresponding underlying field bus networks and be able to operate on this data converting to useful information. The importance of data being transformed into information through complex analytics cannot be underestimated. The ability to have multiple devices and applications that typically have not been able to be connected before all be seamlessly connected for true information integration is so very important.

With all that said I encourage you to read this book cover to cover multiple times. Whether you are an end user or supplier or just someone that wants to understand the manufacturing cycle this book is for you!

Thanks for taking time out from your busy schedules to understand the business and technology of EtherNet/IP.

Thomas J Burke
OPC Founder and Visionary
Cleveland, OH
November 2018

THE EVERYMAN'S GUIDE TO ETHERNET/IP

INTRODUCTION

Everyone loves a good story. We love stories about places, people and things. Even when the story is something as esoteric as EtherNet/IP™. But most of all, we love stories about people. Researchers who study human behavior think this is because it helps us figure out the world and test scenarios in our head. How would I react if my company went out of business? How would I cope with the loss of a loved one or a spouse? How would I manage a major health issue like loss of an arm or leg? How would my life change if I won the lottery? Because what actually happened to other people could happen to us, humans find these kinds of stories very interesting.

One of the most important stories is our "origin" story. We all have an origin story. There is always some key reason that we married our spouse, live in this city, work at this job, were educated at this university and so on. Sometimes it's because a parent guided you in a certain direction. Other times it's because a friend of yours worked at the company, you took a job in a distant city or you simply happened to be in a Starbucks at the same time as your new boss/girlfriend/best friend. There are stories about how a person was seated next to someone on a plane and found a new love, made a career change or received some health advice that saved their life. An origin story tells a lot about someone – what they value, what's important to them and how we can talk to them.

Technologies also have origin stories. Of course, technology origin stories aren't as captivating to most of us as human origin stories are. But often, there is an interesting angle to why a technology developed the way it did and how the circumstances and environment of the time caused that technology to emerge. Techies like us have a lot of interest in these stories.

The EtherNet/IP origin story is fascinating. Where did it come from? Why did it develop as it did and why is it so popular? It's a good story. Not one you would tell your grandchildren around the fireplace, but one you could tell around the bar to other automation guys.

To really understand the EtherNet/IP origin story, we have to go way back. Back to the 1980s and the early days of programmable controllers. Networking in those days was limited except for early adopters of things like RIO, DH+ and Modbus RTU. Eventually it became pretty clear that something new was needed. Process speeds were increasing and there was more and more I/O to transfer (some things never change). There just wasn't enough bandwidth on DH+ and Modbus so something else was needed. Plus, users were tired of proprietary protocols that locked them into a single vendor experience (and the excessive cost structure of that single source solution).

That something else, that new open protocol, turned out to be ControlNet® with a whopping (for the time) five-megabyte data rate. Besides the fact that it was an open protocol, it had precise deterministic behavior. It offered scheduled and unscheduled traffic and it organized I/O and data using an object-oriented structure. That was all unique and novel. It looked to be a real winner.

Unfortunately, what the designers didn't see, and you can't fault them for it, was that we were just about to enter the Ethernet age – an age where Ethernet would come to dominate corporate and industrial communications. Ethernet, with 10 megabaud communications (100 and 1000 meg. now) and standard hardware that everyone soon became familiar with quickly dominated first the business systems, and then, the factory floor.

DeviceNet®, introduced in 1995, was the popular networking alternative, but with the movement toward Ethernet, there was a definite technology gap. DH+ no longer had the required bandwidth; ControlNet cost too much; DeviceNet was too slow with limited bandwidth – so along came EtherNet/IP to fill that gap. With off-the-shelf, low cost, standardized hardware, offering high speed and vastly more bandwidth, it was the right technology at the right time.

And this book is the story of EtherNet/IP: what it is, what makes it special and what you need to know about it.

I'm fairly certain that the EtherNet/IP story won't be a best seller, but I personally think it would make a great movie. But until the movie comes out, you'll just have the book. I hope you enjoy it and learn a little more about EtherNet/IP!

John S. Rinaldi
June 20, 2018
Raleigh, NC

PS: If you like stories, you find more in the Real Time Automation newsletter, the most widely read industrial automation newsletter in the industry. Many manufacturing automation professionals refer to me as "the newsletter guy" and stop what they're doing when it arrives in the good, old fashioned US mail. They get insightful commentary, interesting stories, technical insights, fun, humor and more in my newsletter.

Have a little fun and get some relevant information. Sign up for the Real Time Automation Newsletter!

http://www.rtautomation.com/company/newsletter/

WHAT IS INDUSTRIAL ETHERNET?

Ethernet is the most popular way to form a local area network (LAN) from a group of computers, printers and servers. It is the worldwide de facto standard for linking computers together using a standardized set of electrical and software standards to facilitate the exchange of data. Ethernet is a formal international standard specified as standard 802.3 by the Institute of Electrical and Electronics Engineers (IEEE).

Industrial Ethernet is a specialized, rigorous application of the 802.3 standard applied to the industrial environment that meets one or more of the following requirements:

- **Immunity to Electrical Noise:** Industrial Ethernet communication must operate continually and reliably despite electro-magnetic interference from high-current 480 VAC power lines, reactive loads, radios, motor drives, and high-voltage switchgear.

- **Extreme Environment Ready**: Industrial Ethernet equipment is often located outdoors or in a factory in close proximity to industrial ovens, chillers, freezers and other heat/cold generating equipment.

- **Resistance to Vibration**: Industrial Ethernet devices are often mounted on machinery that moves, shakes, rotates or accelerates.

- **Continuous Operation**: Manufacturing processes often operate 24/7 in critical processes where downtime costs hundreds of thousands if not millions of dollars.

- **Interoperability**: Industrial Ethernet devices must interoperate with very large number of other Industrial Ethernet devices and

with machine controllers, enterprise applications and Cloud applications from vendors like Amazon, Microsoft, GE and Oracle.
- **Determinism**: Industrial systems often require real-time, deterministic responses to machine events.
- **High Performance**: Often, Industrial Ethernet devices must meet application requirements for behavior under erratic machine conditions.

INDUSTRIAL ETHERNET AND TCP/IP

Ethernet 802.3 specifies the electrical and link layer standards for connecting computers but by itself fails to offer the complete solution. Network protocols – systems for transferring data – are needed to make it truly useful. The standard that evolved alongside Ethernet is TCP/IP.

TCP/IP is a suite of communication protocols designed in the 1970s by two DARPA scientists: Vint Cerf and Bob Kahn – persons most often called the fathers of the Internet. It was a development arising out of a project to establish a ground-based radio packet network. That project convinced Cerf and Kahn of the need to develop an open-architecture network model where any computer could communicate with any other, independent of individual hardware and software configuration.

In the early versions of the technology, the core protocol – and there was only one – was named TCP. And in fact, these letters didn't even stand for what they do today, "Transmission Control Protocol," but at that time they stood for "Transmission Control Program." The first version of this predecessor of modern TCP was written in 1973, then revised and formally documented in RFC 675, Specification of Internet Transmission Control Program, from December 1974.

The original TCP expanded from that single transmission protocol to a large suite of protocols that provide seamless network connectivity between computers across the world. Today, TCP/IP, as it's now known, is composed of the Internet protocol (IP), the HyperText Transfer Protocol (HTTP) used by your browser, Simple Mail Transfer Protocol (SMTP) and a host of other protocols that we use countless times a day.

Both from a historical perspective as well as in today's industrial world, the TCP/IP plus Ethernet marriage is a key combination. Neither would have survived or prospered without the other.

DATA ENCAPSULATION

You can think of TCP/IP as a series of pipes that connect senders and receivers. Physical pipes can carry anything and simply don't care if they are carrying water, oil or beer. It's the same for protocol pipes – they don't care what the bits are – but it's very important for the sender and receiver to agree on how to interpret those bits: if the end of a physical pipe is expecting beer and receives motor oil, someone will be very confused (and unhappy).

Senders and receivers agree on how to interpret the bits transferred from a sender to a receiver using application layer protocols like EtherNet/IP, PROFINET IO and Modbus TCP™. TCP/IP provides the infrastructure (the pipe) that encapsulates packets of these protocols and moves them from a sender to a receiver.[1] The protocols in the TCP/IP suite have various capabilities for sending critical data, intermittent data and continuous data.

Practical Implementation of EtherNet/IP

Even though EtherNet/IP is implemented using standard Ethernet, that doesn't mean that all the cables and connectors of standard Ethernet meet the requirements of the factory floor. EtherNet/IP devices are installed in dirty, hazardous, dry, wet, cold, hot, humid and, occasionally, in pristine lab environments. These environments do not affect the signaling or interoperability with standard Ethernet but there are special considerations for these environments.

Cat 5e and Cat 6 are recommended for EtherNet/IP and general Industrial Ethernet cabling applications. Shielded cables are preferred in situations where signal wires are going to be placed in close proximity to devices and cables with large switching currents. In applications with cable runs over 100 m (the limit for most Cat 5e and Cat 6 cables), fiber optic cable is recommended. Fiber optic media are also recommended in applications with very high electromagnetic disturbances.

For a complete discussion of all the considerations for planning your EtherNet/IP network, you should consult the EtherNet/IP Media Planning and Installation Manual from the ODVA. That guide uses the MICE (Mechanical, Ingress, Climatic and Electromagnetic) classification system and provides very detailed information on using EtherNet/IP in more challenging application environments.

---

[1] There is much more on TCP and EtherNet/IP in a later chapter.

THE EVERYMAN'S GUIDE TO ETHERNET/IP

ETHERNET/IP HISTORY

Everyone has one of those seminal moments from their childhood that they'll remember forever. Mine was the 1962 launch of the Mercury capsule that made John Glenn the first human to orbit the earth. It's my oldest memory and one of my most cherished. With the paltry media sources of 1961, the entire world focused on his *Friendship 7* capsule and watched everything: the launch at Cape Canaveral, three orbits of the earth and splashdown in the North Atlantic. What we all didn't know at the time was that a sensor indicated possible damage to his heat shield. NASA thought it was a very real possibility that the first man in orbit would become the first to be incinerated in space. Certainly, a seminal moment for me and for many others at the time.

It wasn't nearly as big, or as momentous, but the mid-1990s launch of DeviceNet also stands out in my memory. Prior to that day, I was a serial guy: UARTS, bytes, RS232, RS485 and all the rest were my thing. In those days, all that you had to know was what goes in on one side and what goes out on the other. We called them the *guzzintos* and *guzzoutas*. We had no concept that it was the stone age of computer networking without a Raquel Welch.[2]

I and many of my colleagues didn't know it at the time, but that DeviceNet launch was my first introduction to CIP, the Common Industrial Protocol. We also didn't know that our professional lives had just changed forever. Without our knowledge, we had just become manufacturing network guys. Instead of guzzintos and guzzoutas, we were managing objects, connections, connection classes, endpoints, message classes, expected packet rates and much more. It was a lot more complicated but extremely exciting. It was as if we went from

---

[2] If you don't get that reference, you weren't a young boy in the late 1960s. You'd have to watch the movie *One Million Years B.C.* to understand.

being earth bound to orbiting the earth.

DeviceNet changed everything. Prior to DeviceNet, almost everything was a custom and proprietary implementation. Modbus existed but it wasn't as popular in those days. If you had a device with a serial port, you designed your own protocol. Usually it was something using an ASCII string. Something like a "$A 0100" to start the process and a "$Z 0200" to stop the process. And sometimes the protocol required you to terminate the character sequences with a carriage return, sometimes with a linefeed and sometimes with both. On the really advanced systems you had your choice!

Matching baud rates between devices in those days was a real headache. If you didn't know the baud rate, data bits and parity of a device (and you usually didn't) you would try all the combinations one by one. Tedious and time-consuming. And very frustrating.

DeviceNet and CAN (Controller Area Networking) solved a lot of those kinds of problems. CAN was new then too. It was developed by Bosch in Germany to solve a huge problem at VW and other German car manufacturers: cost. Remember, the cars of the 1950s and 1960s didn't have power windows, power seats, defrosters or moveable mirrors. If you wanted to adjust the passenger mirror, you parked the car, opened the passenger window and reached through the open window and pulled it left or right (yes, it really was the stone age).

As more and more electronics were put into cars, all the wiring became a massive problem. How can power and control signals be delivered to all these motors all over the automobile? Those wires must be purchased, installed, terminated and fastened in place. Lots of time and labor, and the problem would only get worse in the future.

CAN solved that problem. And since American manufacturing had the exact same problem on the new factory machinery it was developing with more and more fancy electronics, CAN (and DeviceNet) solved that problem for American manufacturing too.

CAN provided the mechanism to easily move messages around from device to device. CAN was, and still is, a link layer protocol. It's really good at moving a small amount of data from one node on a network to another node on the network. The CAN messages contain an eight-byte payload of data. The only responsibility of CAN is to make sure that payload is moved correctly. The payload, of course, was the DeviceNet application layer message. Using two protocols (with one embedded within the other one) – well that was another first for me and many of the geeks like me. Little did we know what the future would bring.

THE EVERYMAN'S GUIDE TO ETHERNET/IP

ETHERNET HISTORY

Around the same time that DeviceNet was under development, there was a lot of activity and competition to be the LAN (Local Area Network) solution for home and business computing. It's hard to imagine today, but it wasn't at all clear in the 1980s and early in the 1990s that Ethernet would be the standard for computer networking. In fact, technologies like Token Bus, Token Ring, and ARCNET each had significant support and backing from an industry goliath promoting that specific technology as the LAN standard.

IBM, for example, supported Token Ring and standardized on it as the LAN solution for its computers. This was at a time when IBM was unarguably the leading technology company in the world. It proclaimed Token Ring as an open standard but, as with many single vendor-driven standards, it was less open than it seemed. Often, non-IBM Token Ring equipment would fail to connect with IBM computers.

ARCNET (Attached Resource Computer NETwork) was the first widely available LAN for microcomputers and became popular in the 1980s for office automation tasks. Introduced in 1977, just five years later there were over ten thousand ARCNET installations. Designed using a revolutionary token passing system, ARCNET was the first "loosely-coupled" LAN solution. It made no assumptions about what kind of device might be at the other end of the link. Over time though, ARCNET was unable to match the speed of other competitive solutions, and ARCNET systems began to vanish in the 1990s though some systems remain in use today.

There was fierce competition between these technologies and Ethernet. Ethernet originated at Xerox Palo Alto Research Center (PARC) in the mid-1970s. The basic philosophy was that any station could send a message at any time, and the recipient had to acknowledge successful receipt of the message. Ethernet had several things going for it over other competing technologies. First, as Urs Von Burg describes in his book, *The Triumph of Ethernet*, DEC decided to support Ethernet. DEC was a leading technology company of the time and very active in the IEEE standards process. With DEC's support, its position as a truly open standard and with more and more developers working on it, Ethernet became the LAN of choice for both the business and home market.[3]

Ethernet was cost-competitive, open and reliable, but the

---

[3] Remember there was no internet at that time.

21

introduction of 10BaseT in 1990 may be the factor that culminated in its acceptance as the LAN standard. 10BaseT allowed the use of hubs and switches. This freed Ethernet from its often cumbersome bus architecture and offered the flexibility of star topology. This change made it much easier for network administrators to manage their networks and gave users far more flexibility in placing their PCs. By the mid-1990s, 10BaseT Ethernet was also much cheaper than Token Ring, no matter which metric you used.

AND CIP IS BORN...

While Ethernet was on its way to becoming the standard for LAN communications around the office and home, a group of manufacturing control vendors was hard at work on a standard more applicable to the factory floor. This group believed Ethernet inadequate for the factory floor as it lacked the repeatability and determinism required in factory floor applications. Carrier Sense Multiple Access (CSMA), where Ethernet devices repeatedly attempt to access the network until they find an open network slot, was deemed completely inadequate for factory floor networking.

ControlNet, introduced in the mid-1990s, combines aspects of an I/O network with a peer-to-peer data network. It offered high speed performance for that era (5 Mbps) and highly repeatable data transfer. With a unique Physical Layer based on RG-6 coaxial cable with BNC connectors, ControlNet had built-in support for fully redundant cabling and messaging that operated in a highly scheduled and deterministic way.

But the high cost of the ControlNet board level interface, higher cost cabling and what turned out to be a lower data rate doomed it. The growing popularity of Ethernet, the introduction of switches that made network segmentation possible, the higher available speeds and the price pressures brought about by the huge consumer and office markets doomed it. Only its object-oriented representation and messaging survived as CIP. In 2001, the basics of CIP were used to form EtherNet/IP, an application layer on top of Ethernet and TCP/IP.[4]

In 2008, ControlNet International, the owner of the ControlNet intellectual property, transferred all rights to ControlNet to the Open Device Vendor Association, who now jointly markets ControlNet, EtherNet/IP and DeviceNet.

---

[4] You'll find much more on CIP and EtherNet/IP in later chapters of this book.

CIP & EIP: TEN THINGS TO KNOW

Most of us in manufacturing are pretty simple and straightforward. Steak with a baked potato rather than oysters, crêpe or pâté. An old pickup instead of a Mini Cooper. Regular old American football, not soccer.

The most simple and straightforward way to learn a new technology is to start with a shortcut – a Reader's Digest[5] of the most important concepts. Something that's a good starting point – a base that you could build from to become proficient.

This chapter is that starting point – a list of the ten most important CIP and EtherNet/IP concepts. If all you require is a quick introduction to the technology, sort of a "speed dating" overview, this chapter is for you.

1. **CIP IS OBJECT MODELING / MESSAGING WITHOUT A TRANSPORT MECHANISM**

    CIP, the Common Industrial Protocol, is a well-defined data representation, connection management and messaging protocol that can operate over many Transport and Physical Layers. CIP defines how devices connect to one another, how they represent data to external entities and how data moves from device to device.

    CIP is an object-based technology that exposes data over a CIP network as a collection of attributes (the data values) grouped in categories called objects. Objects representing common entities with common characteristics can be organized into classes. Actual implementation of a class entity is known as an instance of that class.

---

[5] A small, printed magazine with very short interesting articles, tip, tools and recipes. Sort of a combination of Wikipedia and Pinterest for the pre-internet age.

For example, a pneumatic valve might implement a valve class where each instance represents one of the valves in a valve block.

CIP identifies objects, instances, attributes and the services provided by those objects by integer Identifiers. Two types of connections are defined; Explicit and Implicit. Explicit Message connections are used for transferring non-control data while Implicit Messaging is used for transferring control data.

2. ETHERNET/IP IS A CIP TECHNOLOGY

The Common Industrial Protocol (CIP) does not include Transport, Encoding or Physical Layer communications. The implementation of CIP over some Transport and Physical Layer provides that low level transport. DeviceNet is the CIP implementation over CAN (Controller Area Networking). ControlNet is the CIP implementation over a ControlNet Physical Layer. EtherNet/IP is an implementation of CIP over Ethernet TCP/IP.

3. ETHERNET/IP USES TCP AND UDP

EtherNet/IP messaging makes use of both TCP (Transmission Control Protocol) for critical transfer of non-control information and UDP (User Datagram Protocol) for highly efficient transfer of I/O data.

The Transport Control Layer (TCP) provides connection-oriented, reliable transmission between two computers. TCP handles the establishment of a TCP connection between the two computers, the sequencing of packets, the acknowledgment that the packets were sent, the recovery of packets lost during transmission and the re-establishment of a lost connection.

EtherNet/IP Scanners use TCP to send Explicit Messages. Messages to open a connection, close a connection and read or write an attribute in the Object Model are all done using the connection-oriented and reliable TCP service. TCP was selected for these critical operations as it provides an acknowledgement of delivery.

The User Datagram Protocol (UDP), on the other hand, provides a "fire and forget," connectionless communication service with no acknowledgement of packet delivery. UDP is preferred for small data transfers, where a lost packet is not consequential, or an upper layer protocol is managing reliability. UDP requires less overhead than TCP as it does not need to do connection establishment before transmitting data or provide acknowledgements.

EtherNet/IP Scanners use the UDP protocol for I/O Messaging.

THE EVERYMAN'S GUIDE TO ETHERNET/IP

I/O messages in EtherNet/IP are almost always transferred cyclically. In most cases, the loss of a single message in a cyclical sequence of messages is not critical since another one being transferred on the next I/O cycle, so the small overhead, non-connection oriented UDP service is perfect for I/O Messaging.

4. SCANNERS CREATE CONNECTIONS AND SCAN I/O

We all very familiar with technologies that have some superior/subordinate relationship. Modbus is one that everyone knows. There is a Modbus RTU Master and a Modbus RTU Slave. A Modbus TCP Client and a Modbus TCP Server. A BACnet Client and BACnet Server. It's the same for EtherNet/IP. Only with EtherNet/IP, the terms are different. \ The device that opens connections sends outputs to end devices, and collects inputs from end devices, is a Scanner.

The difference between these technologies and EtherNet/IP is that in the familiar Industrial and Building Automation protocols, the client or master somehow takes ownership of the server or slave device. In many of these technologies, once a client takes ownership of a slave, no other client or master can access it.

That's not true of EtherNet/IP Scanners and Adapters. In EtherNet/IP, an adapter can be configured to accept connections with one, two or any number of scanners. A scanner device can connect and access the data in any number of servers. It's a less exclusive relationship in EtherNet/IP, though, like other technologies, servers simply respond to requests from scanners and never initiate communications. There is, however, one limitation. Only one scanner can control the physical outputs of an adapter. Other scanners can monitor the inputs of an adapter but only one can provide output data.

The key to an EtherNet/IP Scanner is how it is configured. Some scanners have a fixed list of adapter connections while others have various mechanisms for building a scan list of adapters. A later chapter describes the various ways scanners can be configured.

5. ETHERNET/IP ADAPTERS ARE END-POINT DEVICES

An EtherNet/IP Adapter is the end-point or slave side of an EtherNet/IP connection. An adapter measures physical properties, indicates status, initiates physical actions and performs all sorts of physical measurement and activations in the real world and reports them to one or more EtherNet/IP Scanners, usually programmable controllers. EtherNet/IP Adapters are where the physical world meets the digital world.

EtherNet/IP Adapters support two message connections, Explicit Message connections and Implicit Message connections. Explicit Messages are service oriented and use request/response messaging. Implicit messaging use cyclic messaging where scanners cyclically send outputs to adapters and adapters cyclically send inputs to a scanner.

The specific capabilities of an EtherNet/IP Adapter are described by its application objects, the connections it supports and the scope of services it supports. An EtherNet/IP Adapters Object Model represents the data of a device as a set of objects with attributes. Attributes are the data values contained in the object.

6. ETHERNET/IP USES STANDARD ETHERNET

EtherNet/IP uses both the physical and software standards common to IEEE 802.3 Ethernet. In that way, it is identical to any business Ethernet system. EtherNet/IP is simply another application that operates on Ethernet. Just like HTTP, which provides the messaging your internet browser uses, EtherNet/IP is another application that moves messages over Ethernet.

What's generally different about an Industrial Ethernet technology like EtherNet/IP is the physical hardware used to implement it. Industrial Ethernet is a specialized, rigorous application of standard "off-the-shelf Ethernet" technology (IEEE 802.3) that offers additional electrical immunity, more rigorous environmental standards, vibration protection and other specialized mechanisms to allow devices to function in more extreme environments then are generally found in business systems.

7. ETHERNET/IP IS THE ROCKWELL AUTOMATION STANDARD

Rockwell Automation is the largest supplier and promoter of EtherNet/IP devices. The Rockwell Automation Logix programmable controllers provide scanner capabilities while many of its end-point devices such as drives, and I/O blocks provide adapter capabilities. Rockwell's commitment to CIP and EtherNet/IP and their large presence in the US automation market have made EtherNet/IP the North American standard for industrial networking.

8. ODVA MANAGES AND SUPPORTS ETHERNET/IP

The Open Device Vendor Association was founded in 1995 and is now known simply as the ODVA. It is a global association of companies with an interest in the Common Industrial Protocol (CIP).

THE EVERYMAN'S GUIDE TO ETHERNET/IP

ODVA promotes CIP, the CIP derivative technologies and open, interoperable information and communication standards in industrial automation.

ODVA is the legal owner of CIP technology and the CIP network adaptations: ControlNet, DeviceNet and EtherNet/IP. Vendors wishing to incorporate CIP technologies into their process must license the technology from the ODVA. This usually means joining the vendor association and purchasing a technology specification. Membership is only open to vendors actively developing with the technology and not users of the technology.

9. ETHERNET/IP IS A CERTIFIABLE STANDARD

A vendor that licenses EtherNet/IP technology agrees when they sign the ODVA license agreement to obtain, and maintain, a Declaration of Conformity for their product(s). This provides assurance to the users of EtherNet/IP technology that products made and sold using ODVA technologies comply with ODVA's specifications and can interoperate in systems with products from multiple vendors.

10. ETHERNET/IP IS AN I/O PROTOCOL NOT AN IOT PROTOCOL

It's unclear where the idea that Ethernet I/O protocols could be used as IoT protocols started, but the facts are clear. EtherNet/IP, PROFINET and, to a lesser extent, Modbus TCP are perfect as they are. They do a great job of moving I/O data from end devices into controllers. Good enough that it's likely that a majority of the world's manufacturing systems use these protocols.

But there are now some misguided proponents of these technologies that want to use EtherNet/IP, PROFINET and Modbus TCP to move IoT data. That's an indefensible position. Those technologies move I/O data that's very strictly defined between an end device and a controller. That's not what you want in the IoT world for reasons described in a later chapter.

WHAT IS CIP?

WHAT'S INTERESTING?

CIP, the Common Industrial Protocol, organizes and shares data in a way that is completely independent of Transport, Media Access and Physical Layer communications. There is no other manufacturing technology with the openness, structure and organization that can be as easily ported to another field network as CIP.

WHAT DO YOU NEED TO KNOW?

1. CIP is a well-defined data representation, connection management and messaging protocol that operates over some independent Transport and Physical Layers. CIP defines how devices connect to one another, how they represent data to external entities and how data moves from device to device.
2. CIP is the first object-based technology that exposes data over a network as a collection of attributes grouped in common categories called objects.
3. EtherNet/IP, DeviceNet and ControlNet are implementations of CIP that operate over Ethernet, CAN and the ControlNet Physical Layer, respectively.
4. There are two common device types. I/O adapter devices support both Explicit and Implicit operations. These devices are used in simple I/O applications like I/O muxes, photo eyes and valves. Most DeviceNet and EtherNet/IP devices fit this category. I/O scanner devices include everything offered in the I/O adapter but add the capabilities to open connections and initiate message transfers. They can serve as both an originator of output data and a target for devices that want to send them input data.

5. There are several extensions to the CIP Standard including CIP Sync, CIP Motion, CIP Safety and CIP Energy.

WHAT ARE THE DETAILS?

In the early days of automation networking, it was common for manufacturing systems to have a combination of proprietary vendor networks and some home-grown application networks that served some specific purpose for that factory. Since everything was serial and only required a UART (Universal Asynchronous Receive Transmit) for operation, everybody could and did create their own technology for moving control data around the factory.

These systems were not only not open but considered trade secrets by the companies offering these networks. For example, knowledge of the register location of the chilled water temperature in a chiller required not only applying for partnership with the chiller company but signing confidentiality and non-disclosure forms.

As you might imagine, that was a nightmare for creating integrated systems on the factory floor. Most manufacturers had to select a single vendor for the entire system. Using that vendor's proprietary technology was the only way to ensure that every component of your manufacturing system could work with every other component. This, of course, led to exorbitant prices, locked in vendors, heavy support costs and little access to the best technologies offered by smaller companies with innovative ideas and technologies.

GM recognized this in the 1980s and early 1990s. They set out to break the monopoly of single vendor systems and when DeviceNet, the first CIP protocol, was introduced in 1994, they finally had a chance to use the best no matter what company offered it.

DeviceNet led to ControlNet and then EtherNet/IP. It was realized that a true open platform manufacturing system was possible by standardizing the messaging and data representation to create a system that was fully media independent, and CIP was born.

CIP, the Common Industrial Protocol, is a mechanism for organizing and sharing data in industrial devices. CIP is the core technology behind EtherNet/IP, DeviceNet and ControlNet. CIP provides both a common data organization and common messaging to solve various kinds of manufacturing application problems.

CIP Organization

CIP can actually be defined very simply. It is a well-defined data representation, connection management and messaging protocol operating over some independent Transport and Physical Layer (Figure 1).

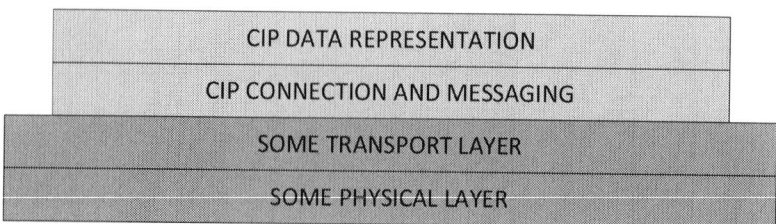

Figure 1 – CIP Overview

The **CIP Data Representation** defines how CIP devices represent data. CIP is an object-based technology, and data exposed over a CIP network is presented as a collection of attributes grouped in common categories called objects. A set of objects can be further grouped into objects for specific applications called application profiles. A subsequent chapter describes the CIP data representation in much greater detail.

CIP Connection and Messaging defines the connection management and messaging that all CIP protocols use. CIP defines specific objects that are used to manage connections and the connection types that specify how data moves over those connections. There are two connection types available in all CIP protocols: Explicit and Implicit. Explicit Message connections are used for transferring non-control data while Implicit Messaging is used for transferring control data. A subsequent chapter describes the CIP connection management and messaging in much greater detail.

CIP does not define how the data bytes of a message physically move from one device to another. CIP is completely independent from the Transport, Media Access and Physical Layers that are used to move messages. In principle, any Physical Layer including RS232 serial communications could be used to implement CIP communications between two devices. However, each CIP implementation specifies a specific Physical Layer, Media Access and Transport Layer. DeviceNet uses CAN (Controller Area Networking). EtherNet/IP uses TCP/IP and Ethernet. ControlNet uses custom, ControlNet specific Transport and Media Access.

Figure 2 presents these layers in more detail. DeviceNet, ControlNet and EtherNet/IP all use the same CIP application layers with different Transport, Media Access and Physical Layers. Any other Physical Layer could also be used to send CIP messages.

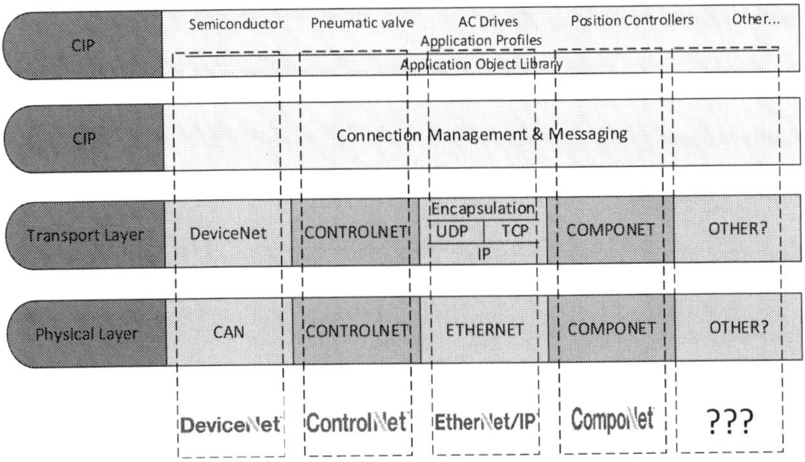

THE EVERYMAN'S GUIDE TO ETHERNET/IP

Figure 2 – CIP Protocol Layers

CIP Device Types

Among CIP technology professionals, there is an unofficial way to classify CIP device types as to their overall functionality (Table 1).

DEVICE TYPE	DESCRIPTION
Messaging Server	Messaging server devices only support Explicit Messaging operations. This kind of device might be used to move ASCII data that occurs intermittently in a barcode or RFID application.
I/O Adapter	IO adapter devices support both Explicit and Implicit operations. These devices are used in simple I/O applications like I/O muxes, photo eyes and valves. Most DeviceNet and EtherNet/IP devices fit this category.
Messaging Client	Messaging client devices only support Explicit Messaging operations but can both initiate and respond to Explicit Messages. This kind of device might be used to pull ASCII data from barcode or RFID applications using Explicit Messaging.
I/O Scanner	I/O scanner devices include everything offered in the I/O adapter but adds the capabilities to open connections and initiate message transfers. They can serve as both originators of output data and targets for devices that want to send them output data.

Table 1 – CIP Device Types

JOHN S. RINALDI

CIP Important Terms

CIP, like all other technologies, has specific terminology related to its operation that all users should know (Table 2).

TERM	DESCRIPTION
Adapter	A CIP-enabled end-device on an EtherNet/IP network. An adapter converts digital signals received from an EtherNet/IP controller to real world outputs and digitizes physical signals and passes those signals as inputs to an EtherNet/IP controller.
Attribute	A data element or parameter within a CIP object. Attributes are addressed by an attribute ID.
Attribute ID	An unsigned integer that indexes the attribute with a CIP object.
Class	An object prototype for a set of objects that share elements, services and other properties. An implementation of the Class is an object instance.
DeviceNet™	An industrial network application layer protocol that utilizes CIP and CAN (Controller Area Network) as the Transport and Networking Layer communications.
EtherNet/IP™	An industrial network application layer protocol that utilizes Ethernet, TCP/IP and CIP to communicate between scanner and adapters.
Explicit Message	A CIP Connection type in which a message is directed to a target node and a response is returned to the message originator. It is used to asynchronously access data in CIP devices.
Input Assembly	An Input Assembly is a block of contiguous data that is assembled from virtual and real-world physical inputs in a CIP end-device and cyclically transmitted to a controller.
Instance	An instance is one, specific realization of an object. An object instance inherits all the properties of the Object Class from which it was derived.
Output Assembly	An Output Assembly is a block of contiguous data that is cyclically transmitted to a controller to set virtual and real-world physical outputs in a CIP end-device.
S	A CIP-enabled master device that originates EtherNet/IP connections, transfers outputs to end devices (adapters) and receives inputs from end devices (adapters).
Service Code	The Service Code of an Explicit Message is a unique code that identifies the service that the Originator is requesting from the adapter. Set Attribute Single and Get Attribute Single are the most commonly used service codes.

Table 2 - Important CIP Terminology

CIP Extensions

There are several extensions to the CIP Standard that while not directly applicable to the subject of this book require mentioning here.

CIP Sync™ provides a mechanism where control applications can coordinate operation to achieve real-time reactions to an external stimulus. CIP Sync™ relies on the IEEE-1588™ standard to achieve synchronization between two devices of fewer than 100 nanoseconds. For EtherNet/IP, CIP Sync is an extension to it into the real-time domain.

CIP Safety™ provides the ability to mix standard CIP devices and safety devices on the same network for increased application flexibility. CIP Safety™ provides fail-safe communication between nodes designed for safe operation such as safety interlock switches, PLCs, light curtains and I/O blocks.

CIP Energy™ provides a set of objects and services designed to optimize energy usage and provides services that allows systems to manage and monitor energy information.

CIP Motion™ delivers an open, high bandwidth, high performance solution for multi-axis, distributed motion control.

CIP OBJECT MODELING

WHAT'S INTERESTING?

If you're new to device data models or your only experience with how embedded industrial devices store data is Modbus, then CIP Object Modeling is going to astonish you. The Common Industrial Protocol (CIP) has a very extensive and sophisticated mechanism for modeling data in a very standard way and making that data available to external devices. CIP provides the identical modeling mechanism in all CIP technologies. Administrative, configuration and diagnostic tools can all access CIP devices in a common and standard way no matter what the underlying transport technology or media that the technology might be using.

WHAT DO YOU NEED TO KNOW?

Here's a quick summary of what you need to know about data modeling in CIP:

1. Other industrial devices use address spaces to describe how data is represented, but CIP uses Object Models to describe how the device designer organizes and represents the data the device is making available to external devices.

2. CIP Object Models are independent of transport and media. The same basic Object Model can be used in DeviceNet, ControlNet and EtherNet/IP.

3. The most elemental entity in a CIP Object Model is an attribute. Attributes are simply data values defined by a name, a numeric identifier and a data type from the list of supported types in the CIP specification.

4. Attributes having some common relationship are grouped together in an Object Class. An object is nothing more than a set

of related attribute values, some services that operate on those attributes and, possibly, some device-specific behaviors.

5. An Object Instance is a specific implementation of an Object Class. There can be any number of Object Instances for an Object Class. Each occurrence of an Object Class is known as an instance of the class and has the identical set of attributes, services and, optionally, behaviors.[6] For example, a Valve Object Class may have an instance for each of the valves in the device.

6. Objects, instances and attributes are identified by numeric identifiers. CIP specifies the allowed numeric ranges for object IDs, instance IDs and attribute IDs.

7. CIP specifies three types of objects: Required objects that must exist in all CIP devices, Application objects that encapsulate data common to lots of devices (organized in the CIP object library) and vendor-specific objects.

8. There is an endless number of ways to organize device data into a CIP Object Model. There is no right way or wrong way to expose data to external devices.

9. CIP offers Device Profiles as a mechanism to provide interchangeability among common device types. Device profiles are Object Models with common application objects, services and state machines that enable users to replace a CIP device from one vendor with an equivalent CIP device from a different vendor.

10. The Assembly object is the most important CIP object. Assembly input and output data attributes in a target device contain the raw input and output data that is exchanged with a CIP originator device.

WHAT ARE THE DETAILS?

The key to understanding the Common Industrial Protocol (CIP) or any communication protocol is understanding how that technology represents data and provides access to that data. Nothing is more important to understanding Modbus TCP, EtherNet/IP, PROFINET I/O, OPC UA or any other technology used on the factory floor. All messaging, configuration, device security, device connections and deployment issues revolve around the data representation.

Representing data in today's industrial devices, the device's address space,[7] now means more than just listing a set of addresses that an

---

[6] This is known in Software Engineering as Inheritance
[7] In devices that represent data as objects, this is also called its object model.

external device can access. Now, as manufacturing becomes more IT-like, there needs to be a unique mechanism to identify the data and a method to categorize it properly, specify its data type and provide access to its metadata representation.[8] Of course, the complexity of the device matters. It's not all that important in a simple, four-cylinder pneumatic valve, but it's very important to a motor drive with hundreds of different kinds of network accessible parameters.

The Address Spaces used by many automation technologies are flat, meaning that there is little to no hierarchical structure to them. Modbus is the best example of a flat Address Space (Figure 3). Modbus has two 64K register spaces and two 64K coil (bit) spaces. The address spaces are composed solely of unsigned integers and binary bits, respectively. There are no organizing elements that provide any additional structure to all those registers and coils.

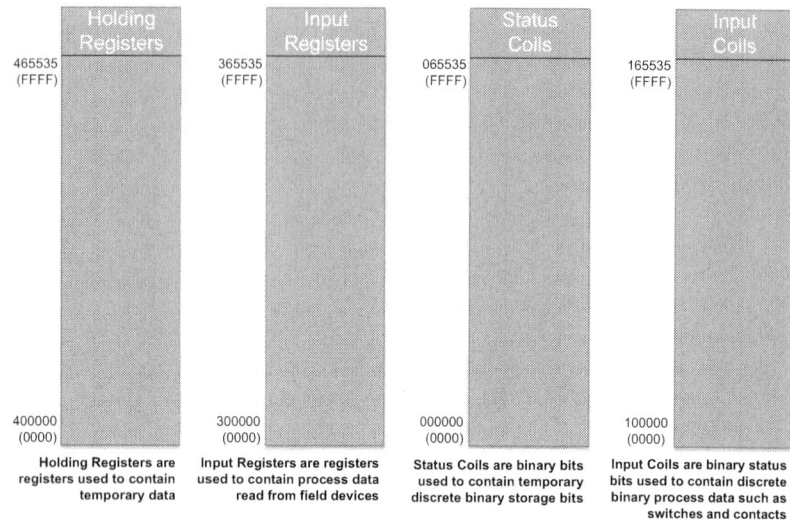

Figure 3 – Modbus Address Space

Newer technologies have vastly improved mechanisms for representing data. OPC UA provides the best example of an open-ended, flexible and hierarchical address space (Figure 4). The OPC UA address space is composed of hierarchical objects. Variables have references to data types, data definitions and an assortment of vendor-defined meta-variables that further classify the variable. In Figure 4, the variable OvenTemp has a metadata variable attached to it called MaxTemp. OPC UA Clients (scanners in CIP terms) can browse

---

[8] Metadata is data that describes and provides additional information about other data.

THE EVERYMAN'S GUIDE TO ETHERNET/IP

through that open-ended, flexible address space in multiple different ways to learn what specific data is available.

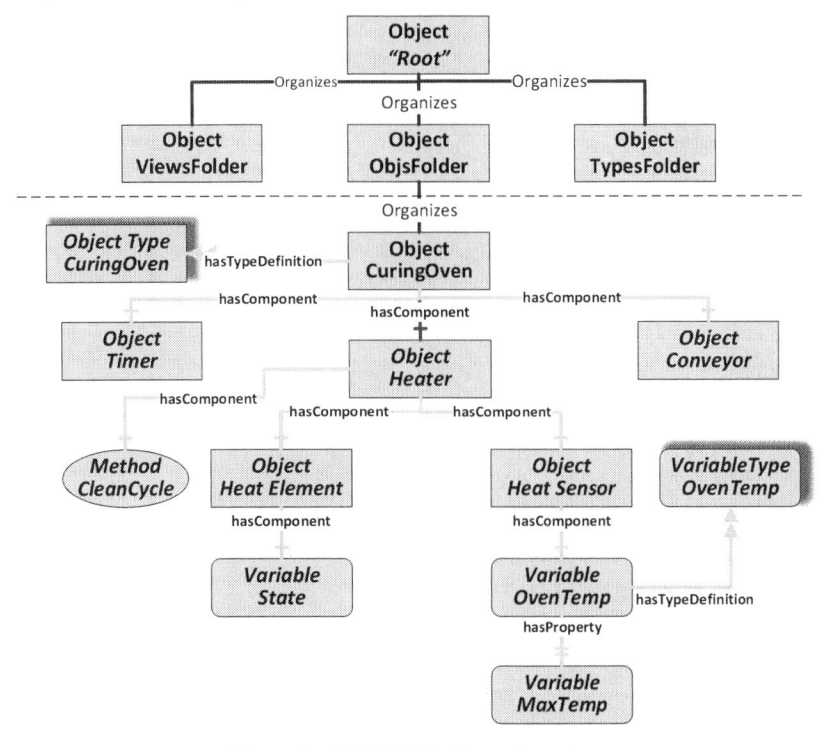

Figure 4 – OPC UA Address Space

The CIP Object Model, the CIP term for a CIP Address Space, is neither as limited as Modbus nor as flexible and hierarchical as the OPC UA Address Space of Figure 4. It does provide mechanisms to organize data, provide standard data types for that data, organize its data for various types of message access and standardize that access for the various CIP technologies (EtherNet/IP, DeviceNet and ControlNet). The base unit or organizing element of a CIP Address Space is the object.

CIP Objects and Attributes

If you can remember your high school chemistry class, you probably remember learning that each element in the periodic table is a substance that cannot be decomposed into other elements. In CIP, that base elemental structure is an attribute. An attribute is simply an element of data in a CIP device. Attributes are defined by the CIP specification, by trade groups or by a device vendor.

Attributes defined in the CIP specification including standard items that appear in all CIP devices like the device's vendor ID, the produced connection size, the expected packet rate or the software revision number. Other attributes are defined by trade associations when they create profiles of standard devices. There are profiles for AC Motors, Position Controllers and other devices. And attributes that communicate some vendor-specific functionality of a device can also be defined by the product vendor.

All attributes are defined as having one of the specific data types defined in the CIP specification. CIP supports a limited, but adequate, set of data types that include integer, floating point, bit, array and other standard data types.

External devices refer to attributes by an attribute identifier, a unique numeric value that identifies a particular attribute. The attribute identifier is used to identify the attribute in operation requests like GET ATTRIBUTE or SET ATTRIBUTE.

All CIP devices (EtherNet/IP, DeviceNet and ControlNet) organize attributes into Object Classes (Figure 5). An Object Class is nothing more than a set of related attribute values. Like attributes, Object Classes are indexed by a numeric Object Class identifier, a numeric value that uniquely identifies an Object Class.

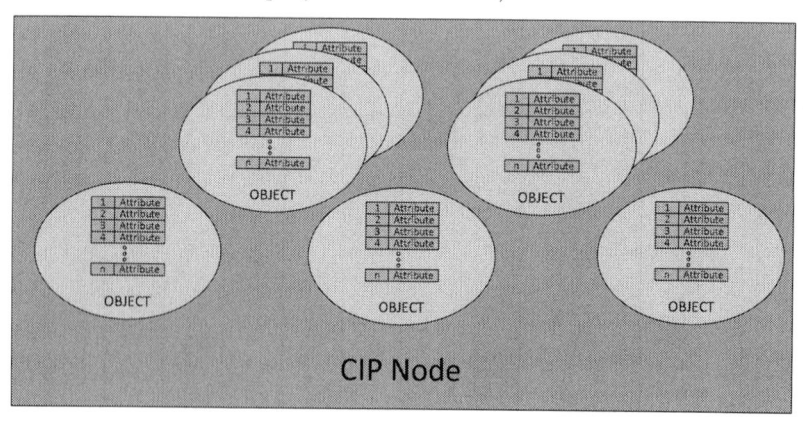

Figure 5 – Structure of Node with Objects

CIP uses Object Classes to describe the content and structure of the data within a node, the services it has available and the behavior of the node. Every CIP device is modeled (but not necessarily implemented) as a collection of Object Classes with attributes, services and sometimes a specific behavior. Any data in a CIP device that is not included in the Object Model is not available to external devices on the network using CIP.

An Object Instance is an actual representation of a single

occurrence of an Object Class entity. Most Object Class entities have a single instance. For example, there is only one instance of the Identity Object Class. In a device with multiple Object Instances, the Object Instances derived from the Object Class (inheritance in software engineering terms) are identical, with the same attributes and services.

In Figure 6, a valve is defined with an Object Class called Valve with two attributes: Current Status and Activation Count. In this example, an implementation of the valve exists with two class objects not organized and four instances of the Object Class Valve. Each Object Instance of the Valve Object Class contains the identical set of attributes: Current Status and Activation Count.

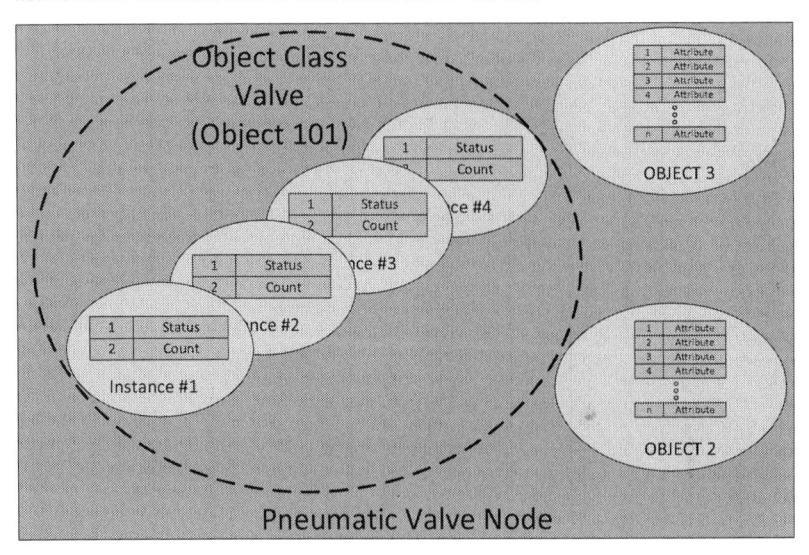

Figure 6 – CIP Object Class Example

Instances, like objects and attributes, are also referred to by a numeric index called an Instance Identifier or Instance ID. Attributes are then fully referenced by an Object Class ID, an Instance ID and an Attribute ID. In Figure 6, the Activation Count of instance 3 is referred to by the reference Object Class ID 101, Instance ID 3, Attribute ID 2.

The Object Class Identifier, Object Identifier and Attribute Identifier are important mechanisms for identifying specific entities in CIP messages. The CIP specification defines how these identifiers are used. Sets of identifiers are assigned predefined values for standard CIP objects while other identifiers are available for use by vendors creating CIP devices. Predefined identifiers indicate entities that are defined in the CIP specification. Table 3 lists the identifier ranges used in CIP devices. Values not shown are reserved.

	Predefined (hex)	Vendor-specific (hex)
Object Class ID	0x0 to 0x63	0x64 to 0xc7
	0Xf0 to 0x2FF	0x300 TO 0x4FF
Instance ID	0x1 to 0x63	0x64 to 0xc7
	0X100 to 0x2FF	0x300 TO 0x4FF
ATTRIBUTE ID	0x1 to 0x63	0x64 to 0xc7

Table 3 – Object, Instance, Attribute Identifiers

CIP Objects

The CIP specification defines three types of objects:

- **Required Objects** – these objects must be included all CIP devices. They provide commonality to all CIP devices. These objects make it possible for network tools to access common attribute data, like the vendor who created the device, in a standard way no matter what the device is or which implementation of CIP (EtherNet/IP, ControlNet or DeviceNet) is used. These objects are listed as "Required Object" in Table 4.

- **Application Objects** – these objects encapsulate the device-specific data. These objects can be objects from the CIP library or other objects described by a predefined Device Profile (see below).

- **Vendor-specific Objects** – these objects describe device-specific data and services that are unique to the vendor. For example, if a vendor implemented a temperature monitor and created a Temperature History object unique to that vendor and not in common use by all other CIP temperature monitors, that would be a vendor-specific object. Data in these objects is strictly of the vendor's choosing and is organized in whatever method makes sense to the device vendor.

THE EVERYMAN'S GUIDE TO ETHERNET/IP

IDENTITY OBJECT CLASS 0x01	REQUIRED OBJECT
The Identify object provides the Identifying information for the CIP Node and includes the vendor id, the product code, the software revision information, the serial number and the product name among other items. The Identify object is a required object and there is usually only a single instance.	
MESSAGE ROUTER OBJECT CLASS 0x02	REQUIRED OBJECT
The Message Router object provides a mechanism for external devices to access objects in a CIP device. Messages sent over Explicit connections are directed to the target object by the Message Router object.	
CONNECTION OBJECT CLASS 0x05	REQUIRED OBJECT
The Connection object is where the characteristics of a connection are maintained in a CIP device. An instance of the Connection object is generated for every connection. That instance identifies the connection as Explicit or Implicit, sets the packet rate on Implicit connections and holds other descriptive information on the connection. A connection object is removed when the connection is closed.	
ASSEMBLY OBJECT CLASS 0x04	OPTIONAL
The Assembly object provides the interface to CIP devices communicating with a device over an Implicit connection. Instances of an Assembly object organize the data that is exchanged with external devices. An Input Assembly instance organizes the data that is transferred to external devices. An Output Assembly instance organizes the data that is transferred from external devices. Multiple Assembly instances may be defined, and an external device can choose (by Instance ID) which Assembly instance to use in the transfer.	
PARAMETER OBJECT CLASS 0x0F	OPTIONAL
A Parameter object provides a standard mechanism for a CIP device to make its configuration parameters publicly available to external devices. It provides complete identifying information for the configuration parameters of a CIP device.	
NETWORK SPECIFIC LINK CLASS 0x*NN*	REQUIRED OBJECT
The network specific Link object provides the information on the specific link (DeviceNet, EtherNet/IP, ControlNet) used to implement the CIP device. The object specifies attributes that describe the link, such as the node addresses and data rates. See the Chapter on EtherNet/IP operation over CIP for more details on the Link object for EtherNet/IP.	
APPLICATION OBJECT CLASS IDs 0x64 to 0Xc7	OPTIONAL
Application objects organize the specific data and services of a device. Vendors building CIP devices can choose to implement no application objects, one application object with all the data for a device or any number of application objects.	

Table 4 – Some Common CIP Objects

The Object Library

In addition to the objects listed in Table 4, the CIP specification defines a long list of application objects that CIP device manufacturers can implement. A few of these are listed in Table 5. These objects provide a common way for vendors of CIP devices to organize standard data.

CLASS CODE	CLASS NAME
07 hex	Register
08 hex	Discrete Input Point
09 hex	Discrete Output Point
0A hex	Analog Input Point
0B hex	Analog Output Point
0E hex	Presence Sensing
10 hex	Parameter Group
12 hex	Group
1D hex	Discrete Input Group
1E hex	Discrete Output Group
1F hex	Discrete Group
20 hex	Analog Input Group
21 hex	Analog Output Group

Table 5 – The CIP Object Library (Partial)

CIP Object Modeling

The Object Model of a CIP device reflects how the device designer wants to organize the data in the device. There is no right way or wrong way to organize data. For example, if you had a CIP flow meter that measured two flows and two temperatures, you could organize your data in a multitude of ways including:

- As a single Flow Meter Object Class containing four attributes: two attributes for each flow value and two attributes for each temperature value.
- As two objects classes: a Temperature Object Class containing two attributes for the two temperatures and a Flow Object Class containing attributes for the two flow rates.
- As two objects classes: a Flow 1 Object Class and a Flow 2 Object Class. Each Object Class contains two attributes: a temperature and a flow.

CIP Device Profiles

To provide interchangeability between CIP devices, the CIP

THE EVERYMAN'S GUIDE TO ETHERNET/IP

specification includes a feature called Device Profiles. Device profiles ensure that identical devices from different manufacturers contain the identical base Object Model, configuration and behavior, allowing users of the devices to easily replace a device from one vendor with a product from a competing vendor with a minimum of configuration.

A Device Profile defines the required and optional Object Classes to include, state machine behaviors, how I/O data is formatted, the configuration parameters used, and much more. Device profiles are categorized by device types and provide consistency between similar devices of the same device type.

Supporting a Device Profile allows end users to easily integrate common devices. Motor drives are the best example of this. All major motor drive vendors support the AC Motor drive profile. A controller configured to connect to a motor drive with that profile can use any drive supporting the series of objects defined in the profile with minimal reconfiguration. There are Device Profiles (libraries of application objects) for drive systems, motion controllers, valve transducers, and more.

The Assembly Object

One of the most important CIP objects and one of the least understood by novices is the Assembly Object Class.

Data within Application Layer objects are grouped into Assembly Object Classes. An instance of an Assembly Object Class contains data from one or more application or vendor-specific objects. Assembly Object Instances form the data packets that are exchanged by an adapter and a scanner. The data attribute of an Input Assembly Object Instance contains the input data delivered to a scanner. The data attribute of an Output Assembly Object contains the latest output data a scanner delivered to a CIP adapter device.

Assemblies are defined by the product vendor and can be composed as desired by the device vendor. Vendors can support multiple, different assemblies and allow a scanner device to choose an assembly appropriate for a specific application. For example, a vendor of a Temperature Controller with multiple temperature loops may define an assembly for each temperature loop and an assembly with data from all temperature loops. The controller can then pick the assembly that is most suited for the application.

Assemblies are usually predefined by the vendor, but CIP also includes a mechanism by which the user can dynamically create an assembly from Application Layer Object attributes. Few, if any, devices implement that mechanism.

CIP MESSAGING

WHAT'S INTERESTING?

What's interesting about the CIP messaging architecture is how CIP simplifies connection establishment, connection management and connection messaging. CIP predefines the transports, encoding and much of the message connection management. Predefining the interaction between CIP devices simplifies protocol development and streamlines the operation and ease of understanding of CIP communications.

WHAT DO YOU NEED TO KNOW?

There are only two message connection types available in CIP: explicit connections and implicit connections. Explicit connections are request/response oriented and used to asynchronously access data in an adapter device. Implicit connections (I/O messages) are used for control and status. Inputs cyclically flow from an adapter to a scanner. Outputs cyclically flow from the scanner to the adapter. Explicit Messages, used to transfer non-control data, use a TCP connection while Implicit Messages, used to transfer inputs and outputs, use a UDP connection.

Explicit Messages explicitly identify the command and data in the message packet. Explicit Messages are used to read attributes of CIP objects, write attributes of CIP objects, initiate services and perform other operations both predefined and vendor-specific. Implicit messages, on the other hand, contain data that is implicitly understood by the CIP devices using the implicit connection. Both must have some prior understanding of the contents of the message.

TCP (Transmission Control Protocol), which provides message acknowledgement, is always used for explicit connections. CIP devices communicating using Explicit Message connections typically transfer operational commands, and confirmation of those messages is sometimes critical to the sender. TCP provides that confirmation.

UDP (User Datagram Protocol) is used for Implicit connections. Since Implicit Messages are cyclically and continuously exchanged over implicit connections, confirmation of individual messages is not required. The low overhead, connectionless, fire-and-forget nature of the UDP protocol is more efficient for Implicit Messaging.

WHAT ARE THE DETAILS?

CIP messaging is one of the three primary components of the Common Industrial Protocol. CIP messaging is the part of CIP that specifies how ControlNet, EtherNet/IP and DeviceNet open, manage and maintain connections and how messages are constructed and transmitted over those connections.

CIP is connection-based. All data transfer between two CIP nodes occurs over a connection (Figure 7). On the transmission side, there is always an application object that generates a message and a communication object that provides the outgoing interface. On the receiving side, there is always a communication object that receives the message and an application object that processes it.

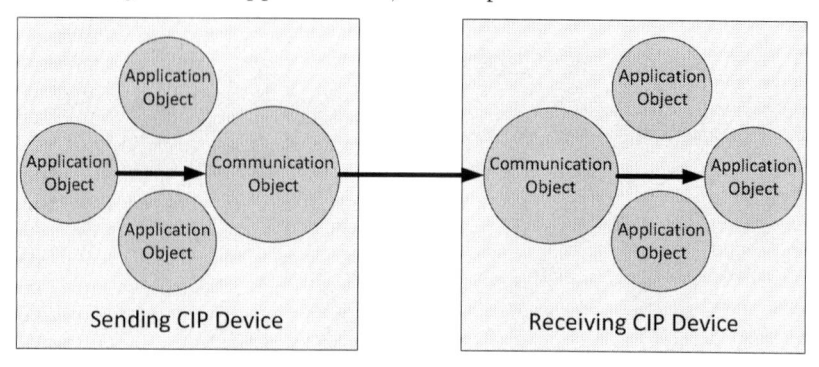

Figure 7 – CIP Communication Objects

CIP Messaging is described by the Messaging Connection objects, the connection establishment process, the connection paths that are used to route message and the two types of CIP messages.

CIP Messaging Connection Objects

There are only two types of CIP messages: Explicit Messages and Implicit Messages. Both message types are established using the same connection object, but the resulting connections operate differently.

THE EVERYMAN'S GUIDE TO ETHERNET/IP

In an Explicit Message connection, there is a unicast path between the producing device and the consuming device (Figure 8). A producing device has some application object that generates the Explicit Message request. A consuming device has an explicit connection that receives the message request and routes it through the router object for processing. The router identifies the correct application object for processing the request and routes the message to that object. The process is reversed for the response to the Explicit Message request.

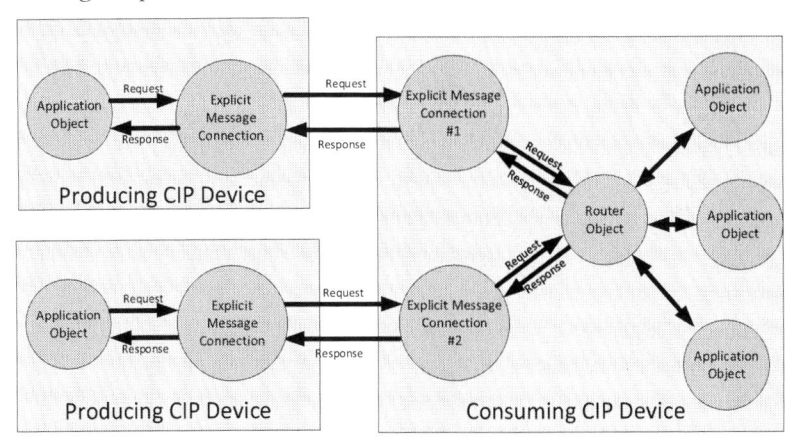

Figure 8 – Communication Objects for Two Explicit Connections

In Figure 8, the consuming device has two Explicit Message connections, each of which is connected to a different producing device. The number of connections that a device can support is device dependent and usually limited by available resources.

The communications objects for Implicit Messaging are quite different (Figure 9). All implicit communications begin and end with an Assembly object. An Input Assembly object collects attributes from application objects and produces them on an implicit connection. An Output Assembly object essentially distributes attributes to application objects when it is received by the Implicit Messaging connection. A CIP originating device, like a controller, generally has many implicit connections. In Figure 9, CIP Device #1 has a producing and consuming connection with CIP Device #2 but only a listen-only connection with CIP Device #3.

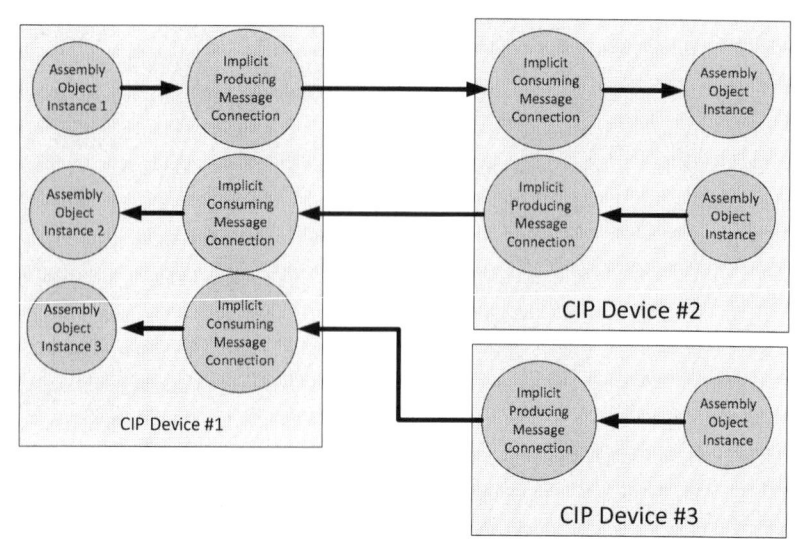

Figure 9 – Implicit Messaging Communication Objects

In both Explicit and Implicit connections, there are two parts to the communication structure: the producing/consuming application object and a communication connection object. In both, the attributes of the Connection object define how the connection operates. Some of the more important attributes of a Connection object are listed in Table 6.

ATTRIBUTE	DESCRIPTION
Connection State	An indicator of the current state of the communication object. Valid possible states are non-existent, established, timeout among others.
Instance Type	An indicator for whether an instance specifies an explicit or implicit connection.
Transport Class Trigger	This attribute indicates the direction of the message, how the message is produced and the message class.
Produced Connection Size	The number of bytes produced by the connection, or zero if a consuming only connection.
Consumed Connection Size	The number of bytes consumed by the connection, or zero if a producing only connection.
Expected Packet Rate	The frequency at which messages are expected on this connection

Produced Connection Path	Set of logical segments specifying the application objects whose data will be produced by this connection.
Consumed Connection Path	Set of logical segments specifying the application objects whose data will be consumed by this connection.

Table 6 – Important Attributes of the Connection Object Class

The interface between a Connection object and an application object consists of a series of logical segments. A logical segment is a series of data bytes that identifies an application object. There are a number of ways defined in the CIP specification for structuring logical segments, but they are typically identified by as a series of identifier/data pairs as shown in Figure 10.

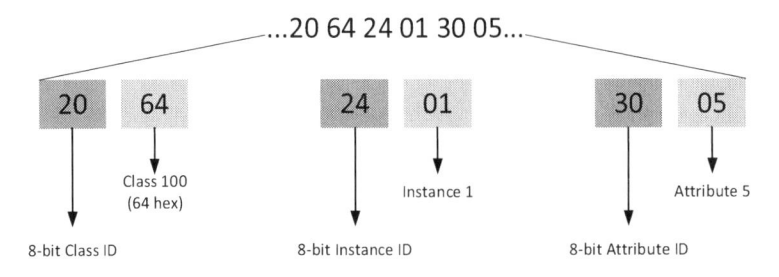

```
          ...20 64 24 01 30 05...
```

20	64		24	01		30	05
	Class 100 (64 hex)			Instance 1			Attribute 5
8-bit Class ID			8-bit Instance ID			8-bit Attribute ID	

Figure 10 – Logical Segment Example (Hex Values)

Logical segment addressing is used in various places by CIP devices. It can be found in Electronic Data Sheet (EDS) files, in parameter identification, in unconnected messaging and other places.

CIP EXPLICIT MESSAGING

Explicit Messages (Figure 11 and Figure 12) use the traditional request/response sequence to access data or initiate operations in a target device. Message originators usually (but not always) use Explicit Messages to transfer non-control data to a target device. More infrequently, message originators use Explicit Messages to read and write I/O data. In either case, message targets respond by sending explicit response messages back to the message originators.

In all CIP protocols, the master (or scanner) device is usually the message originator and a slave (or adapter) device is the target device. Sometimes though, a device can have both scanner and adapter functionality. In that case, it can function as either the message originator or message target.

Explicit Messages are termed "explicit" because the format of the request and the response are explicit and well-known. Each message contains some sort of function or service code, a path to an object in an address space of the CIP device and any optional data associated with the service code. A listener intercepting an Explicit Message can easily open the message and decode it as you can see in the following figures.

CIP EXPLICIT MESSAGE PACKET

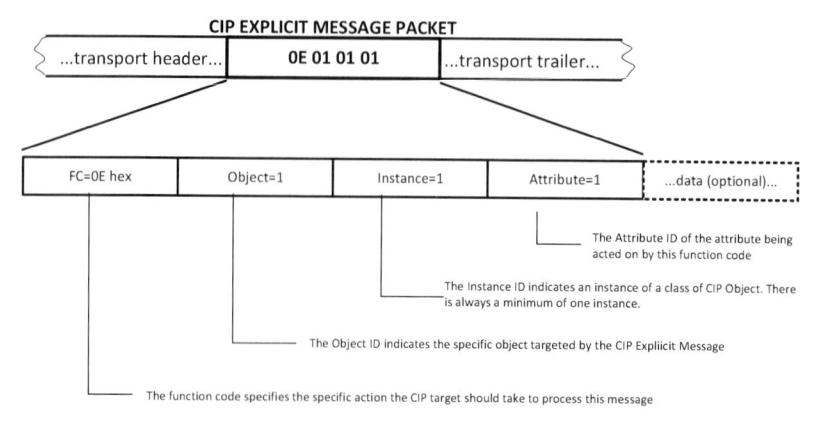

...transport header... | 0E 01 01 01 | ...transport trailer...

| FC=0E hex | Object=1 | Instance=1 | Attribute=1 | ...data (optional)... |

The Attribute ID of the attribute being acted on by this function code

The Instance ID indicates an instance of a class of CIP Object. There is always a minimum of one instance.

The Object ID indicates the specific object targeted by the CIP Expliicit Message

The function code specifies the specific action the CIP target should take to process this message

Figure 11 – Explicit Message Packet Contents

Explicit response messages use a similar message format to the original Explicit Message. The response contains a response code with the original function code with the high order bit set, a status code identifying the result of the operation and any data reported as a result of the requested operation.

CIP EXPLICIT RESPONSE PACKET

...transport header... | 8E 00 00 55 (hex) | ...transport trailer...

| 1xxx xxxx – Response ID | Status=00 | Data Value = 00 55 |

Data (optional) returned by the execution of the original command

The Status code provided by the target device

The Response ID is the original Explicit Response service code with the high bit set

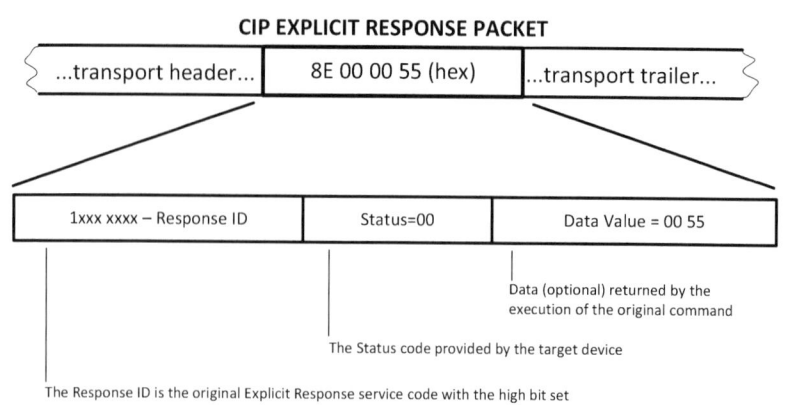

Figure 12 - Explicit Message Response Format

There are a number of Common service codes listed in the CIP specification. Each CIP object supports one or more of these service codes. The CIP specification lists exactly what service codes are supported for each object and how the object processes a service code.

The most common CIP service codes are listed in Table 7. Note that some of these services are only supported on selected objects.

SERVICE CODE	NAME	DESCRIPTION
0E hex	GET ATTRIBUTE SINGLE	Returns the current value of an attribute specified by the object/instance/attribute description
10 hex	SET ATTRIBUTE SINGLE	Sets the current value of an attribute specified by the object/instance/attribute description
01 hex	GET ATTRIBUTE ALL	Returns a list of all the current values for all attributes of an object
05 hex	RESET	Instructs the target to Reset. The particular semantics of a reset are specific to the target object.

Table 7 – Common CIP Service Codes

The CIP Service Codes listed in Table 7 are also applicable to application and vendor-specific objects. The CIP specification specifies what service codes are supported for which application objects. It is up to the device vendor to decide which, if any, of the common service codes should be supported on a vendor-specific object. The vendor can also create unique service codes with vendor-specific functionality for vendor-specific objects. The unique, vendor-specific service codes used by the vendor must conform to the CIP Service code range table listed in Table 8.

SERVICE CODE RANGE	DESCRIPTION
00 – 31 hex	Common Service codes assigned to required and vendor-specific objects
32 – 4A hex	Services created by vendors to implement vendor-specific functionality and assigned to application objects created by that vendor
4B – 63 hex	Services specific to an entire Object Class including instances
64 – 7F hex	Reserved for future use
80 – FF hex	Not Available – Reserved for Reply service codes

Table 8 – Service Code Identifier Ranges

CIP Implicit Messaging

Implicit Messages are termed implicit because the contents of an Implicit Message are implicitly understood by both the originator and the target. The contents of an Implicit Message always refer to an Assembly object, which aggregates attributes of objects in an address space. An input (data item moving from target to originator) Implicit Message contains the data aggregated from attributes in application objects in the target. An output (data item moving from originator to target) Implicit Message contains data to be written to attributes in application objects in the target.

Implicit messages appear to listeners intercepting the messages as unformatted bytes that might be a series of bits, a set of integers, some number of floating point values or anything else. Listeners intercepting Implicit Messages without the knowledge of the message assembly have no means to decode the message. Examples of Implicit Messages are presented in Figure 13 and Figure 14.

| ...transport header... | 04 13 09 12 25 | ...transport trailer... |

Implicit Input Message is simply a string of data bytes

Figure 13 – CIP Implicit Input Message Packet

| ...transport header... | 09 12 15 24 05 04 13 09 12 25 | ...transport trailer... |

Implicit Output Message is simply a string of data bytes

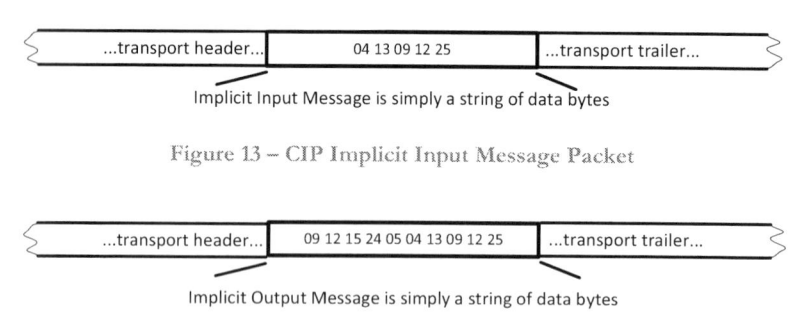

Figure 14 – CIP Implicit Output Message Format

CIP Connection Establishment

CIP defines a mechanism to connect two unconnected devices. The process varies slightly with the particular CIP technology but centers around a service called the Unconnected Message Manager (UCMM). The connection requester sends an UCMM Forward Open request to initiate a connection. The Forward Open request contains everything that is needed for the target to create the connection including but not limited to:

- The identity of the requesting device; the Vendor ID and Serial number attributes of the Identify Object
- The Connection ID (CID) for the connection from the requestor to the target The Connection ID (CID) for the connection from the target back to the requestor. Timing information for the connections; the requested packet interval for each connection
- The Connection path from the Connection object to the application object (I/O Connection)

- The Trigger mechanism that is to be used for the application object to generate data (cyclic or Change of State (CoS)

The UCMM service in the target receives the message, creates the requested connection using the specified connection attributes and returns a status message to the requesting CIP device.

EtherNet/IP and ControlNet make the most use of the Forward Open command. DeviceNet, designed for smaller, less resource-rich devices, makes use of a simplified connection process and rarely, if ever, uses a Forward Open request.

CIP & ETHERNET/IP

WHAT'S INTERESTING?

What's interesting is that many people think that EtherNet/IP is just CIP over an Ethernet connection. They aren't wrong, but there's a lot more to the story. What EtherNet/IP is, in truth, is actually more complicated than just CIP over an Ethernet connection. EtherNet/IP includes a mechanism to create connections and download parameters. It adds additional objects required to monitor and control Ethernet and TCP/IP behavior and it augments CIP messaging with an entire Encapsulation Protocol that can not only encapsulate CIP but other proprietary protocols as well.

WHAT DO YOU NEED TO KNOW?

EtherNet/IP is much more than CIP over Ethernet. This is what you need to know:

1. EtherNet/IP adds two additional objects to the standard CIP Object Model: the Ethernet Link object and the TCP/IP Interface object. The Ethernet Link object provides the representation for physical link information while the TCP Interface object provides a representation of the TCP/IP network setting. Both objects are used to monitor and configure Ethernet and TCP/IP communications.

2. EtherNet/IP includes an Encapsulation Protocol that provides the messaging interface for encapsulating CIP messages and sending them over TCP/IP. Both CIP Explicit and Implicit (I/O) messages are encapsulated by the Encapsulation Protocol to form the messages that are transmitted over TCP and UDP.

3. The EtherNet/IP Encapsulation Protocol can also serve to encapsulate other protocols, proprietary or open (like a Modbus)

THE EVERYMAN'S GUIDE TO ETHERNET/IP

and transfer those protocol messages from one EtherNet/IP node to another.
4. The EtherNet/IP Encapsulation Protocol provides a set of services that allows a device to find other EtherNet/IP devices, register sessions and send data to devices.
5. EtherNet/IP uses both connected and unconnected messaging. Connected messaging is used for continuous, ongoing messaging while the Unconnected Message Manager (UCMM) is used for one-time, sporadic messages.
6. EtherNet/IP uses the Forward Open command via the Unconnected Message Manager (UCMM) to open Explicit and Implicit connections. Forward Open specifies timing, paths to assembly data and the production triggers that initiate message transmission.

WHAT ARE THE DETAILS?

EtherNet/IP is an implementation of CIP, an object-based technology. Data exposed over a CIP network is presented as a collection of attributes grouped in common categories called objects. CIP defines the connection management and messaging that all CIP protocols use (including EtherNet/IP). EtherNet/IP supports the two connection types available in CIP: Explicit and Implicit. Explicit Message connections are used for transferring non-control data while Implicit messaging is used for transferring control data.

But EtherNet/IP is much more than just CIP over Ethernet. EtherNet/IP builds on CIP and defines some additional technology that is specific to operation over Ethernet. EtherNet/IP includes additional objects that define how data specific to Ethernet communications is represented. EtherNet/IP defines an Encapsulation Protocol that defines how Explicit and Implicit messages are formatted as TCP and UDP data packets. EtherNet/IP supports both connected and unconnected messaging and multicast messaging to allow multiple receivers to monitor input data from an EtherNet/IP node.

EtherNet/IP Objects

CIP uses objects to describe the content and structure of the data within a node, the services available in a node and the behavior of the node. Every CIP device is modeled (but not necessarily implemented) as a collection of objects with attributes and services. Any data in a CIP device that is not defined as an attribute in the Object Model is not available to external devices on the network using CIP.[9]

Objects can be grouped into Object Classes. Object Classes are objects that provide a reference structure for a group of objects each of which represents a single instance of some common entity. An Object Instance is an actual representation of a single occurrence of that Object Class entity. Because they reflect a common entity, the object structures of the objects derived from the Object Class are identical with the same attributes and services (i.e., inheritance in software engineering terms).

The CIP adaptation for Ethernet – EtherNet/IP – extends that common object structure used for all nodes to include two additional nodes, the Ethernet Link object and the TCP/IP Interface object (Figure 15).

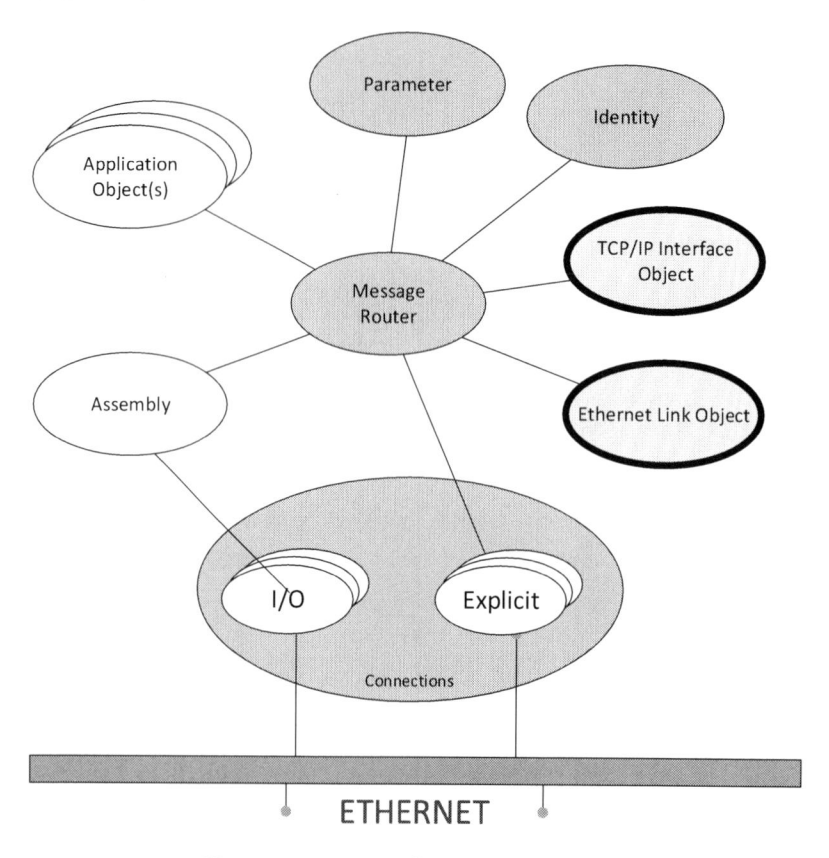

Figure 15 – EtherNet/IP Object Structure

---

[9] An earlier chapter describes the CIP Object Model in much greater detail.

The Ethernet Link object provides the representation for link specific information. Attributes in the Ethernet Link object include the interface speed, Link Status flags and the MAC layer address of the device. The Link Status flags are especially instructive. They indicate if the device is operating in half or full duplex, the status of the link speed negotiation (Auto, Forced, Auto Failure), hardware faults and other important status information. There is generally only a single Ethernet physical interface on most EtherNet/IP devices and one Ethernet Link object to represent that interface.

The TCP/IP Interface object provides a mechanism to represent the TCP/IP network interface. An assortment of attributes is defined for the TCP/IP network interface including interface status, control flags, a physical link path and the TCP/IP address, mask, gateway and name server information. Status information includes its BootP configuration (BootP, DHCP or other), configuration state, configuration of multicast address and much more. The TCP/IP Interface object provides the mechanism for network originators to monitor and/or configure the TCP/IP parameters of a target device.

Encapsulation

Encapsulation is a mechanism for limiting access to data by enclosing it within some other object or mechanism. It is a very common and very important term in computer software but not as common to those who don't write code for a living. In the wider subject of object-oriented programming, it means to restrict access to data by controlling the interfaces that are allowed to manipulate the data. In computer networking, it simply means to enclose a data set, usually a computer protocol, as the data for another computer protocol.

Much of Ethernet and network connectivity in general is built on the concept of encapsulation. The standard OSI model is an encapsulation model. The Transport Layer of the OSI model is designed to be encapsulated within the network layer, which is encapsulated within the Link layer, and so on.

The TCP/IP protocol suite uses encapsulation. Protocols within TCP/IP like ARP (Address Resolution Protocol), HTTP (HyperText Transfer Protocol) and FTP (File Transfer Protocol) are all protocols that are encapsulated by other layers of TCP/IP.[10]

EtherNet/IP also relies on encapsulation. EtherNet/IP encapsulates CIP messages and sends them over UDP or TCP, the

---

[10] An earlier chapter contained a more detailed discussion of CIP and TCP/IP

JOHN S. RINALDI

Internet Protocol (IP) and the Ethernet Physical Layer (Figure 17). CIP Explicit Messages, which generally require delivery acknowledgment, are encapsulated by TCP (Transmission Control Protocol). CIP Implicit messages, which contain cyclic I/O data, are encapsulated in UDP (User Datagram Protocol), the "fire-and-forget" protocol.

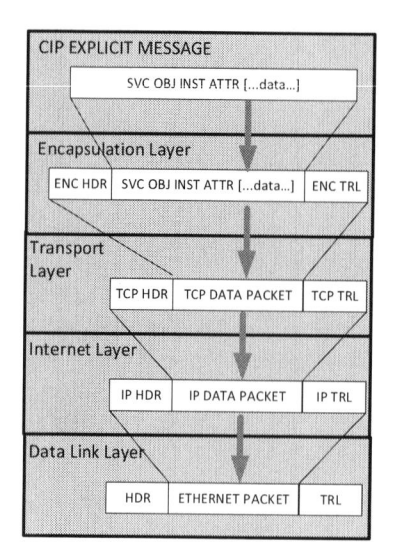

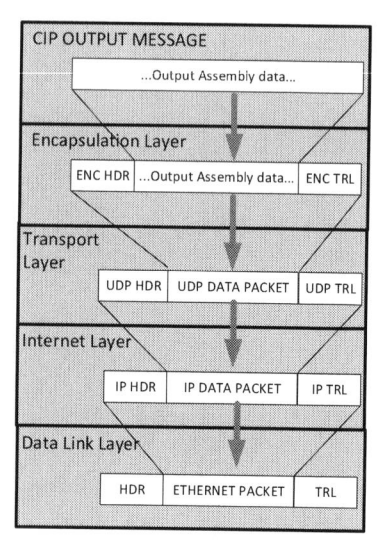

Figure 16 – CIP Explicit and Implicit Message Encapsulation

It's important to understand that encapsulation of CIP Implicit (I/O) messages is different from the encapsulation of CIP Explicit Messages. When an I/O message is encapsulated, the entire message is encapsulated in a single message. TCP, on the other hand, is more of a streaming service. A TCP packet may encapsulate the end of one Explicit Message and the beginning of the next, or it could encapsulate an entire CIP Explicit Message.

Figure 17 presents another, more extensive, view of EtherNet/IP encapsulation. What you'll notice from Figure 17 is that other industrial protocols can be encapsulated in the EtherNet/IP Encapsulation Protocol. You could, for example, embed a Modbus RTU message in the EtherNet/IP Encapsulation Protocol. Or you might embed a proprietary serial protocol message. The EtherNet/IP Encapsulation Protocol is not only for CIP messages.

THE EVERYMAN'S GUIDE TO ETHERNET/IP

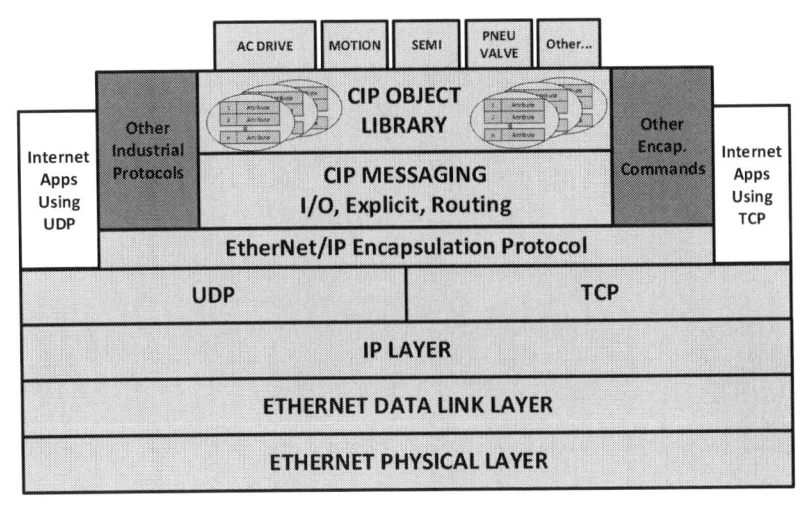

Figure 17 – CIP Encapsulation

EtherNet/IP incorporates a set of encapsulation commands (Table 9) that are not available in other CIP technologies. Some of these commands can only be requested by the originator while others can be requested by either the originator or the target. Most of these encapsulated commands use TCP for message transport but some use the UDP Transport Layer.

One of the most important responsibilities for the EtherNet/IP encapsulation manager is session management. Sessions are nothing more than a protocol binding on top of the TCP connection between an originator and a target. The Encapsulation Protocol manager is responsible for establishing, maintaining and closing sessions. The session management commands are listed in the Encapsulation Protocol command table (Table 9).

COMMAND	DESCRIPTION	SVC
NOP	Sent by the originator or the target to verify operation of a TCP Connection	TCP
ListIdentity	An Originator broadcasts a ListIdentity command to identify potential CIP target devices. CIP devices reply with identifying data that includes attributes of the Identity Object. Software tools are typically the originators of ListIdentity requests.	UDP TCP
ListInterfaces	ListInterfaces is used by Originators to identify non-CIP communication interfaces that could be used by commands including the SendRRData command.	UDP TCP
RegisterSession	An Originator sends a RegisterSession command to initiate a session with a target device. A	TCP

	session handle is returned in the reply.	
UnregisterSession	An originator or a target sends an UnRegiserSession to terminate a session and close the TCP connection.	TCP
ListServices	An originator sends the ListServices command to determine which encapsulation service classes the target device supports. The ListServices reply contains a list of service classes supported but only one is defined, the Communications service class. A reply containing the communications service class indicates that the target device supports encapsulation of CIP messages.	UDP TDP
SendRRData	An originator issues the SendRRData command to transfer an encapsulated request packet to a target. The target issues a SendRRData reply packet to return the encapsulated reply. The actual CIP request/reply packets are encapsulated in the data portion of the message. The SendRRData command is used to send unconnected messages to a target including the Forward Open command.	TCP
SendUnitData	An originator issues the SendUnitData command to transfer an encapsulated request packet to a target. The target issues a SendUnitData reply packet to return the encapsulated reply. The actual CIP request/reply packets are encapsulated in the data portion of the message. The SendUnitData command is used to exchange messages as part of an Explicit connection.	TCP

Table 9 – EtherNet/IP Encapsulation Commands

EtherNet/IP Connected vs. Unconnected Messaging

EtherNet/IP supports connected and unconnected messaging for Explicit Message connections. Connected messaging is messaging across a previously allocated connection. Unconnected messaging is messaging without a previously allocated connection.

Though there are a lot of factors involved, the important differences between the two come down to resources and performance. Continuous messaging generally uses fewer resources when transferred over a connection. There is more overhead to process an unconnected message than a connected message, but there is no point in establishing a connection for messages that are sporadic or one-time. With these messages, there is no reason to spend the resources and processes to open and keep a connection that isn't going to be used frequently.

THE EVERYMAN'S GUIDE TO ETHERNET/IP

Another factor that must be considered for message type selection is the availability of connected and unconnected message ports on the message target. Programmable Controllers often have a mix of connected and unconnected message ports and sometimes less of one and more of the other.

Unconnected messages, by definition, are always used for connection establishment. EtherNet/IP target devices support a message manager called the Unconnected Message Manager (UCMM). This message manager is used in EtherNet/IP and ControlNet to establish new connections. A UCMM Forward Open command is sent as an unconnected message to a target to begin the connection establishment process. The Forward Open command contains important connection related parameters, many of which are used to configure the instance of the Connection Manager object for that connection.

Parameter	Description
Originator Identity	The serial number and vendor ID of the device originating the Forward Open
Connection Timeout	Timeout information for the connection
Connection Type	Information about the connection including whether it will be a multicast or unicast I/O connection
Connection Size	Maximum message size for the connection
Production Trigger	How and when data will be produced by each side of the connection
Requested Packet Interval (RPI)	How often data will be produced on the connection
Connection Path	The path to the Assembly data for I/O connections

Table 10 – Forward Open: Most Important Parameters

To close a connection, the originator sends a Forward Close message. The target must release all the resources associated with the connection, close the TCP connection and inform the application that the connection is no longer active.

EtherNet/IP Multicasting

One hard and fast rule of EtherNet/IP is that only one originator can send outputs to a target. You can imagine the havoc if two EtherNet/IP devices were both to send different speed settings to a motor.

But there are legitimate reasons for multiple originators to want to monitor inputs from a target device. Other devices may be interested in a motor speed – for example, to better manage their operation. Other devices may want to know when a button is pushed or when a pallet arrives at a processing station.

This kind of messaging, where multiple devices see the input data from a device, is called multicasting and it uses the IGMP service of TCP/IP. IGMP is the Internet Group Management Protocol. It provides a mechanism to establish membership among some group of devices on a network. That group of devices joins the group and they can then receive special announcements from other devices on the network.

The router is device providing IGMP on the network. It manages subscription requests from devices and does the heavy lifting of distributing messages to all the devices that have joined the group.

TCP & ETHERNET/IP

WHAT'S INTERESTING?

What's interesting about TCP/IP and its relationship to EtherNet/IP is that EtherNet/IP is simply another application layer protocol that rides on top of TCP/IP. It's just like an email application, a web browser, Yelp or Facebook – an application that needs the services of a TCP/IP stack to operate.

The other interesting aspect of this relationship is that TCP/IP manages all the communications with the Ethernet PHY and the Ethernet MAC. The Physical Layer Interface, the PHY, provides the electrical signaling for Ethernet while the Media Access Controller (MAC) manages the network timing and the access to the network. Because of TCP/IP, EtherNet/IP doesn't need to know anything about the embedded processor being used for the scanner or adapter. EtherNet/IP doesn't need to know the operating system being used or even if the device is using an embedded operating system. It simply needs to know what TCP/IP stack is being used so that it can properly use the protocols of the TCP/IP stack to send messages over Ethernet.

WHAT DO YOU NEED TO KNOW?

There are six important things to know about TCP/IP and EtherNet/IP.

1. A TCP/IP stack is the collection of rules, processes and protocols for sending messages over Ethernet.
2. A TCP/IP stack is layered. The protocol of one layer of a TCP/IP stack is embedded as the data packet of the next layer. Sending a message requires building up the successive protocol layers.

Receiving a message is the process of decomposing the layers until you get the EtherNet/IP application layer message.

3. EtherNet/IP is simply CIP over a TCP/IP stack and Ethernet.
4. EtherNet/IP uses TCP (Transmission Control Protocol) to transfer Explicit Messages.
5. EtherNet/IP uses UDP (User Datagram Protocol) to transfer cyclic I/O messages.
6. The ODVA has specific requirements for selecting a TCP/IP stack to include in an EtherNet/IP device.

WHAT ARE THE DETAILS?

The architecture of an EtherNet/IP Adapter can take many forms. Some vendors choose to embed EtherNet/IP in a hardware add-on, others use a pure software solution and others, heaven help them, build their own. No matter which development path is selected, there must be a TCP/IP stack because EtherNet/IP is an application layer protocol that relies on the protocol layers of a TCP/IP stack to move messages. By definition, EtherNet/IP is CIP communications over TCP/IP and Ethernet.

What is TCP/IP?

TCP/IP is a term that many people use (and misuse), and if you're a networking novice, you may not have a clear idea of what it really is. It's actually very simple. TCP/IP is the collection of communication protocols that are needed for two computers to communicate over Ethernet. It is the set of rules and procedures that have been agreed upon across the world so that any computer can talk to any other computer. It doesn't matter if you are buying socks on Amazon, sending and receiving e-mail or Skyping with your friend Emily in Michigan, you are using a TCP/IP stack like the one described by Figure 18.

THE EVERYMAN'S GUIDE TO ETHERNET/IP

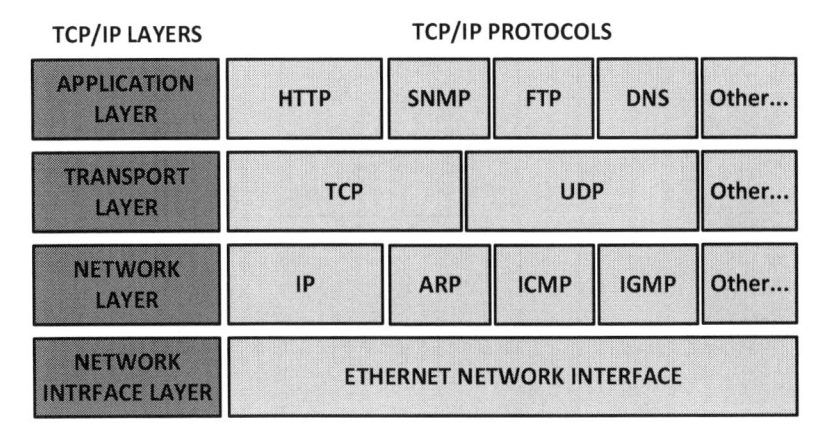

TCP/IP LAYERS	TCP/IP PROTOCOLS				
APPLICATION LAYER	HTTP	SNMP	FTP	DNS	Other...
TRANSPORT LAYER	TCP		UDP		Other...
NETWORK LAYER	IP	ARP	ICMP	IGMP	Other...
NETWORK INTRFACE LAYER	ETHERNET NETWORK INTERFACE				

Figure 18 – Some of The Many TCP/IP Protocol Layers

There are two important items to note about Figure 18. First, you should note that the protocols of a TCP/IP stack are layered. They build upon one another. HTTP, for example, needs the TCP protocol as a transport, IP as a network layer and the hardware interface of the Ethernet Network Interface to bring up your Facebook home page. The second thing you should note is that this figure doesn't show all the protocols of a TCP/IP stack. There are many more than the abridged group shown in Figure 18.

How a TCP/IP Stack Works

Each of the protocols in a TCP/IP stack perform a very specialized task. ARP, for example, the Address Resolution Protocol, is a low-level protocol that maps TCP addresses to the low level, 48-bit hardware address that identifies every Ethernet node. ARP operates by identifying, for example, that the Ethernet node with the address of 192.168.0.100 has a MAC (Media Access Controller) address of 00:eb:24:b2:05:ac (hex). Even though the upper layers of the TCP/IP stack are happy with 192.168.0.100, the Ethernet layer can only send messages to MAC addresses, so ARP provides a very important service.

Service functions, like ARP, perform some ancillary service for other components of the stack. Other components, protocol layers that contribute to messaging across a network, are organized into layers where the protocol message of one layer is embedded as the data of the next layer. This is illustrated in Figure 19.

In this illustration, an HTTP message to request a web page is created by the HTTP application layer on the left. That request becomes part of the TCP message in the next layer. The TCP

Transport Layer becomes the data packet of the IP protocol layer and that message becomes the data packet of the Ethernet message. And when received at the destination, the process is reversed until the application layer message is received by the HTTP application layer at the destination node. It fulfills the request for the web page and returns it by reversing the process to send the response back to the requester.

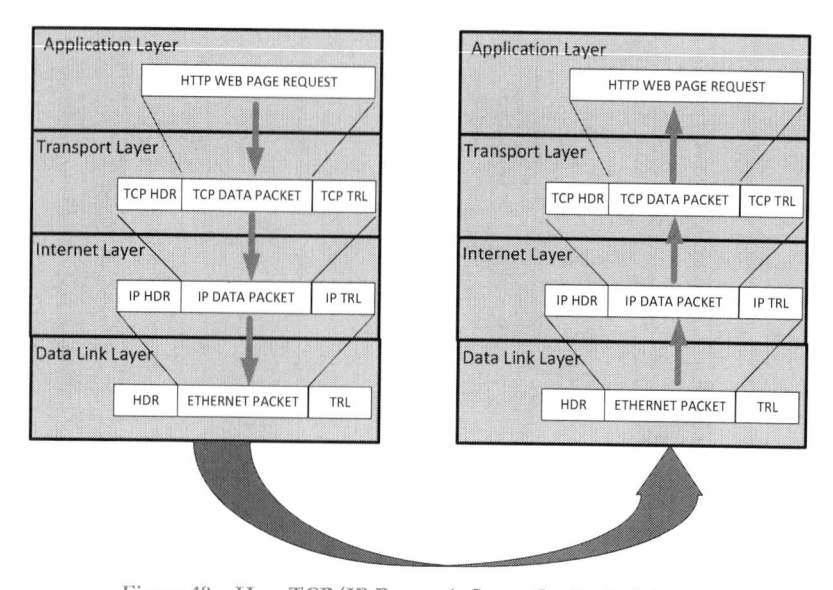

Figure 19 – How TCP/IP Protocols Layer On Each Other

The Important Layers of a TCP/IP Stack

There are several very important layers in the previous illustration that are used by all Ethernet Application Layers including EtherNet/IP and require additional comment.

The **Transmission Control Protocol Layer** (TCP) provides connection-oriented, reliable transmission between two computers. TCP handles the establishment of a TCP connection between two computers, the sequencing of packets, the acknowledgment that the packets were sent, the recovery of packets lost during transmission and the re-establishment of a lost connection.

The **User Datagram Protocol** (UDP) provides a "fire and forget," connectionless communication service with no acknowledgement of packet delivery. UDP is preferred for small data transfers, where a lost packet is not consequential or where an upper layer protocol is managing reliability. UDP requires less overhead than TCP as it does not need to do connection establishment before transmitting data.

The **Internet Protocol Layer** (IP) is a low-level TCP/IP protocol layer that manages moving data from one computer to another computer. It routes the packet by providing the hardware address of the next destination node for the packet and, if necessary, fragments large packets and manages the fragmentation. The IP layer is the lowest software layer in a TCP/IP stack. IP messages are given to the hardware/software that manages transmission over the Ethernet Physical Layer.

EtherNet/IP and TCP/IP Stack

None of the protocol layers discussed in the previous section know anything about CIP or EtherNet/IP. A CIP Explicit Message such as "Read attribute 3 of instance 2 of object 100" is simply an application message embedded in the TCP layer and then the other layers of the TCP/IP stack. A CIP message is no different to the TCP/IP stack than an SMTP (mail messages), HTTP (browser requests) or any other application layer message.

EtherNet/IP uses both TCP and UDP to transfer messages between scanners and adapters. EtherNet/IP Scanners use the connection-oriented TCP layer to send Explicit Messages. Messages to open a connection, close a connection, or read or write an attribute in the Object Model are all done using the connection-oriented and reliable TCP service. TCP is used because it provides message acknowledgement. If a scanner directs an adapter to start Recipe 10, TCP provides the acknowledgement that the message was received by the server.

For I/O Messaging, EtherNet/IP Scanners use the less reliable, fire and forget UDP protocol. I/O messages in EtherNet/IP are almost always transferred cyclically. In most cases, the loss of a single message in a cyclical sequence of messages is not critical since another one is being transferred on the next cycle, so I/O messages can be transferred over the small overhead, non-connection oriented, less reliable UDP service.

Selecting a TCP/IP Stack

The ODVA has specific recommendations for the selection of a TCP/IP stack for an EtherNet/IP Adapter (or a Scanner). All EtherNet/IP devices shall at a minimum support:

- Internet Protocol (IP version 4) (RFC 791)
- User Datagram Protocol (UDP) (RFC 768)

- Transmission Control Protocol (TCP) (RFC 793)
- Address Resolution Protocol (ARP) (RFC 826)
- Internet Control Messaging Protocol (ICMP) (RFC 792)
- IEEE 802.3 (Ethernet) as defined in RFC 894
- Devices receiving Multicast traffic must also support Internet Group Management Protocol (IGMP) (RFC 1112 & 2236)

NOTE: EtherNet/IP devices are encouraged but not required to support other application layer protocols and internet protocols not specified such as HTTP, Telnet, FTP, etc.

THE ETHERNET/IP ADAPTER

WHAT'S INTERESTING?

What's fascinating about EtherNet/IP Adapter devices is the commonality between EtherNet/IP Adapters and DeviceNet and ControlNet Slave devices. Each of these end devices supports common message connections, a common Object Model structure and a common service set. That's very unique to technologies in Industrial Automation and enabled by the Common Industrial Protocol (CIP).

The advantage to the CIP architecture is that a manufacturer can select the CIP technology most appropriate for a particular application. Because of its advantages in systems with linear topologies, DeviceNet might be selected for a conveyor line. And for performance reasons, EtherNet/IP might be selected for machine I/O. But in both cases, both the controllers and end-devices operate in much the same fashion with common Object Models, connections and service sets. That is very unique and a testament to the power of CIP.

Another interesting aspect of EtherNet/IP is flexibility of the Object Model to support any manufacturing application. A device designer can begin with the base of common objects, add objects specific to a particular device type and then add objects that support some vendor-specific function. All these objects- standard or custom - can be accessed in a standardized way using common messaging. Also unique, and a further testament to the power of CIP.

WHAT DO YOU NEED TO KNOW?

EtherNet/IP Adapters provide access to I/O in many different industries in many different applications. Here's a summary of what you really need to know about EtherNet/IP Adapters:

An EtherNet/IP Adapter is an end device on an EtherNet/IP network that measures physical properties, indicates status, initiates physical actions and performs all sorts of physical measurement and activations in the real world under the direction of an EtherNet/IP Scanner – usually a programmable controller. An EtherNet/IP Adapter is where the physical world meets the digital world.

Adapters use the two message connections defined by the Common Industrial Protocol (CIP): Explicit Messaging connections and Implicit Messaging connections. Explicit Messages are service oriented and use request/response messaging. Implicit messaging uses cyclical messaging. In cyclical messaging an EtherNet/IP Scanner cyclically sends outputs to EtherNet/IP Adapters and EtherNet/IP Adapters cyclically send inputs to scanners.

A TCP transport is used for Explicit Messaging connections. A UDP transport is used for Implicit Message connections.

The specific capabilities of an EtherNet/IP Adapter are described by its Object Model, the connections it supports and the scope of services it supports. An EtherNet/IP Adapter's Object Model represents the data of a device as a set of objects with attributes. Attributes are the data values contained in an object.

Objects can have instances, which provide a common look and feel and common services to a set of similar objects. For example, a valve device might define a Valve Class object with specific attributes and services related to the functioning of a pneumatic valve. Each specific valve might be implemented as an instance of the valve class inheriting those same attributes and services from the Valve Class object.

Interchangeability is important to end customers and device manufacturers. Adapter Object Model profiles provide interchangeability in EtherNet/IP. Profiles define a common set of objects and services for a class of objects. The most widely used profile is the AC Motor drive profile, which standardizes how a scanner – typically a programmable controller – accesses an adapter implemented using the AC Motor Device Profile.

The Assembly object is arguably the most important object in an adapter's Object Model. An Assembly object defines the structure of the input data transferred to the scanner and the output data received from the scanner. Input and output data are defined as instances of the Assembly object and usually include specific attributes from other objects of the Object Model.

EtherNet/IP Adapters provide services to scanners over Explicit Message connections. There are common services like Get Attribute Single, Set Attribute Single, Forward Open and Forward Close. There are also services that are specific to classes of objects and vendor-specific services (Start and Stop). Vendor-specific services are unique to that vendor's device. "Load Recipe" is an example of a service that would be specific to particular vendor's device.

WHAT ARE THE DETAILS?

Server devices, and we're talking industrial server devices here, generally provide some sort of digital representation of the physical world. Servers measure physical properties, indicate status, initiate physical actions and do all sorts of physical measurement and activations in the real world under the direction of a remote client device. It's where the physical world meets the digital world.

EtherNet/IP servers are fundamentally just like all other servers in that they are also endpoint devices. Just like other kinds of servers, they measure and digitize inputs and transform outputs to their analog equivalents. They have similar security (or lack of it), similar operation, a similar service set, error checking, Physical Layer and more.

But that is where the similarity with EtherNet/IP ends. First, the terminology is different. EtherNet/IP servers are termed Adapters not servers. And secondly, EtherNet/IP Adapters are much more sophisticated than servers of other technologies like Modbus TCP, Modbus RTU and a host of other similar devices.

EtherNet/IP Adapter devices are end-point devices that are more capable than servers of other industrial technologies. EtherNet/IP Adapters are designed to function in a wide variety of applications from low end sensor/actuator devices to sophisticated instruments and factory floor automation devices. EtherNet/IP Adapters are generally more functional, provide more services, offer superior diagnostic capabilities and provide critical I/O faster and more reliably.

The specific capabilities of an EtherNet/IP Adapter are described by its Object Model, the connections it supports and the scope of services it supports.

EtherNet/IP Adapter Object Model

An EtherNet/IP Object Model[11] provides the external representation for the data the device designer wishes to present to the outside world. All industrial devices implement some sort of address space. The address space is how a device represents its functionality to other devices on the network. The address space is how the inputs, outputs and data in a device are represented. In EtherNet/IP, since the address space is represented by objects, an EtherNet/IP address space is referred to as its Object Model.

The EtherNet/IP Object Model represents data differently than many other technologies. A Modbus address space, for example, consists of a series of 64K blocks. The PROFINET and Profibus address spaces are represented as Racks, Slots, Modules and Channels. Another popular technology groups data into something called topics. There are many ways of representing data in an industrial device, but the EtherNet/IP Object Model represents data in a way that is both understandable and flexible without cumbersome complexity.

The Object Model of a CIP adapter device such as an EtherNet/IP Adapter or DeviceNet Slave is composed of a set of objects with attributes that represent the data elements the device designer wants to make available over the network.

Figure 20 illustrates the general structure of an EtherNet/IP Object Model. The dark gray objects represent the objects that exist in every CIP device; the light gray objects represent objects specific to EtherNet/IP; while the very light gray objects represent the objects specific to the application.

---

[11] An earlier chapter described CIP Object Modeling in great detail.

THE EVERYMAN'S GUIDE TO ETHERNET/IP

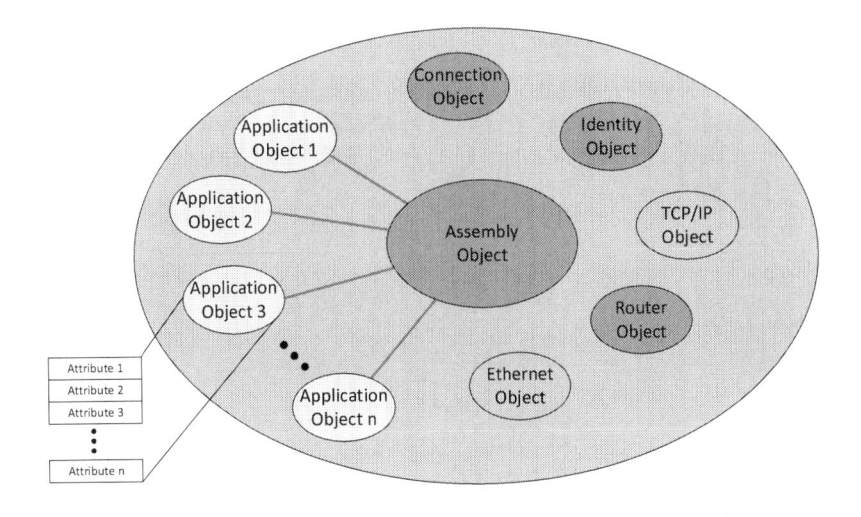

Figure 20 – The EtherNet/IP Adapter Object Model

The attributes of the objects of an Object Model can be accessed by an EtherNet/IP Scanner using either of the connection types available in CIP. Using an Explicit Messaging connection, any attribute in the Object Model can be accessed. Read only attributes are only available for read operations while all other attributes are available for both read and write. Using Implicit Messaging connections, a scanner can access the attributes included in the input and output assemblies. These assemblies and the attributes they contain are then exchanged with a scanner using cyclical communications.

The five most important objects of an EtherNet/IP Adapter are the Assembly object, the Identity Object, the Connection Manager object and the TCP/IP Interface Object..[12]

The **Assembly object** is arguably the most important object in an EtherNet/IP Object Model. Using Implicit Messaging, attributes included in an input instance of the Assembly object are transferred to a scanner and attributes included in an Output Assembly instance are transferred from a scanner to an adapter.

Normally, most adapters support a single Input Assembly and a single Output Assembly. In some application cases, device designers have defined a multitude of Assemblies: everything from only a single Input Assembly to multiple Input Assemblies and Multiple Output Assemblies. This allows a scanner device to pick and choose what set

---

[12] See the chapter on CIP Object Modeling for a more detailed discussion of Object Modeling, Assemblies and standard CIP Objects.

of attributes it wants to cyclically send and receive.

The **Identity object** is the object that identifies the adapter to EtherNet/IP Scanner devices. It contains attributes that include the adapter name, the software revision, the vendor identification and the product name and serial number.

The **TCP/IP Interface object** is the object that provides attributes to monitor and configure the TCP connection. These attributes include all the usual suspects that you need to configure a TCP/IP connection: the IP address, the subnet mask and the gateway IP address.

The **Ethernet Link object** is the object that provides attributes that characterize the Ethernet link. These attributes include the Ethernet link speed and the MAC Address of the EtherNet/IP device.

The **Connection Manager object** is the object that provides the services to open and close explicit and implicit connections. Optional metric parameters are exposed to present device characteristics, like the number of used connections and the maximum bandwidth supported (packets per second).

EtherNet/IP Adapter Message Connections

CIP defines two kinds of connections: Explicit Message connections and Implicit Message connections. Explicit Message connections are used to open connections, close connections, read and write attributes of the Object Model and request services to be performed. Implicit Message connections are used to transfer inputs from the adapter to a scanner and outputs from the scanner to the adapter.

An adapter can support any number of Explicit Message connections. Resource-rich devices can support simultaneous connections to as many scanners as resources permit. Having more than one Explicit Message connection allows a number of tools, scanners or other applications to simultaneously access objects, attributes and services in the adapter device.

However, many EtherNet/IP Adapters are resource constrained devices, and those devices generally support only a single explicit connection. In extremely low resource devices, the explicit connection is closed to free resources after it is used to allocate an Implicit Message connection.

The Explicit Message connection is used to open the Implicit Message connections that are used by the majority of EtherNet/IP

THE EVERYMAN'S GUIDE TO ETHERNET/IP

Adapter devices. An Implicit Message connection is not required. An adapter can be ODVA certified as compliant to the EtherNet/IP specification that only exchanges data over Explicit Message connections. However, that that is unusual and discouraged as most EtherNet/IP Scanners are programmable controllers, and dedicated logic must be coded to issue Explicit Messages. Programmers generally dislike adapter devices that don't support Implicit Messaging.

An adapter can support as many Implicit Message connections as it wants; however, most adapters support only a single connection to conserve resources. More I/O connections means that multiple scanners can access input data, with only one scanner – the first one to connect – having the ability to write the Output Assembly.

Most adapter connections are unicast (one scanner to one adapter) but multicast operation is possible. IGMP (Internet Group Management Protocol) is used for multicast operations. In multicast, the Ethernet switch manages subscribers and publishes the Implicit Messages to all the subscribers on the multicast list. Multicast operation is supported in switches. Typically, there is no special adapter configuration or logic required to support multicast operation.

EtherNet/IP Adapter Services

An EtherNet/IP Adapter can execute services on request from a scanner device. The service requests are transmitted to the adapter over an open Explicit Message connection. The adapter executes the requested service and transmits a response to the service as an explicit response over the same Explicit Message connection.

Some service requests like a "Start" or a "Stop" requires no data. Other services require minimal data like "Start Job," which might simply contain an integer job number. Some services might have extensive data on the service request or the service response. For example, a "Load Recipe" service might require a long and complicated sequence of data. A "Get History" service might return thousands of bytes of history from a Mass Flow Controller..[13] The amount of data attached to a service or a response is fixed for common services and vendor defined for application services and vendor-specific services.

CIP services are attached to objects. Several of the CIP required objects offer services..[14] One of the common, predefined services is

---

[13] "Load Recipe" and "Get History" are examples of vendor-specific commands particular to some vendor's device.
[14] See the CIP specifications for a detailed listing of services offered by each CIP

the Reset service, which is attached to the Identity object and used to execute a soft reset of the EtherNet/IP Adapter device. Vendors typically define services and attach them to application objects where they perform some service specific to that object. A maintenance object might have a service request called "Start Clean Cycle" that begins a cleaning cycle on a device. In a service like that, the response message might simply indicate that the cleaning cycle was initiated. Or it might return an error condition with a response code indicating that the device is in a state where it can't begin a cleaning cycle.

Adapters with services that can be completed immediately will send an immediate response in the explicit response to the initial request. Adapters with services that require seconds, minutes or hours to complete have various ways of communicating status. Some may use the service response to provide current status information. Others might have an alternate service message for status information. Others might indicate status in an attribute that is part of an Input Assembly. And still others might use a combination of techniques to provide service status to a requestor.

EtherNet/IP Adapter – A View Under the Hood

If you want to understand how an EtherNet/IP Adapter is architected, it's not all that complicated. Figure 21 shows the various software layers you might find in an EtherNet/IP Adapter.

---

object.

THE EVERYMAN'S GUIDE TO ETHERNET/IP

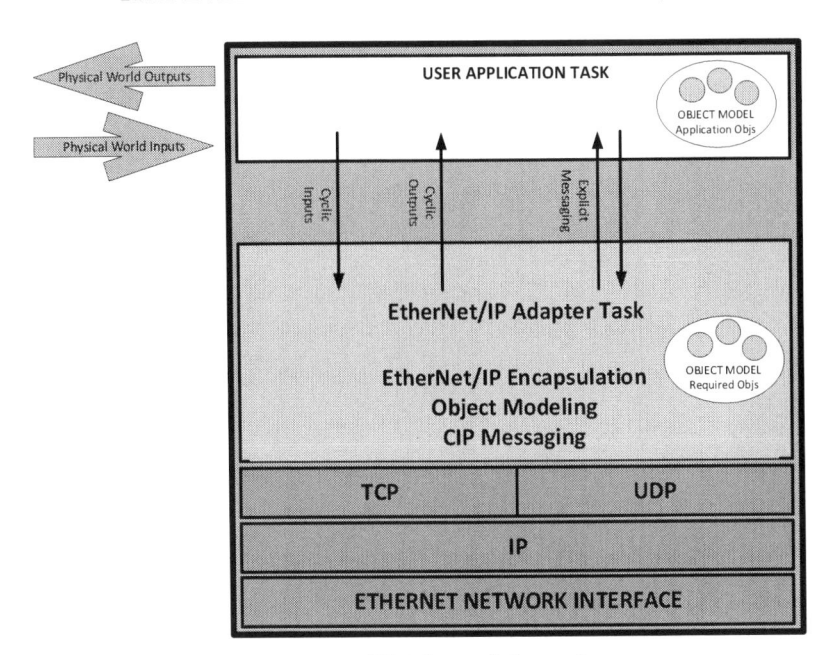

Figure 21 - Ethernet/IP Adapter Software Structure

There are three main layers in this architecture:

1. **The TCP/IP Stack:** the TCP/IP stack (dark gray area) provides the communication protocols needed to move messages across an Ethernet network. These protocols include TCP (Transmission Control Protocol), UDP (User Datagram Protocol), IP (Internet Protocol) and the Ethernet physical interface. The TCP and UDP layers provide transports for moving messages between an EtherNet/IP Adapter and an EtherNet/IP Scanner. The IP and Ethernet Physical Interface provide low-level routing of messages between the nodes of the Ethernet network.

2. **The EtherNet/IP Adapter Software:** the EtherNet/IP Adapter software layer (gray areas) provides both CIP (the Common Industrial Protocol) and the EtherNet/IP encapsulation layer. This layer is where CIP Explicit and Implicit Messages are formed and encapsulated in the Encapsulation Protocol. An Output Assembly received from an EtherNet/IP Scanner is decoded and provided to the user application when a cyclic implicit output message is received. Input Assembly data from a user application is encoded and used to form implicit input messages to cyclically send to (an) EtherNet/IP Scanner(s). In addition, this layer also decodes Explicit Messages received from a scanner and passes the service code and its data, if any, to the user application. Responses generated by the user application to those service messages are

encoded and returned to the originator of the original Explicit Message.

3. **The User Application:** The user application is the software that implements the basic functionality of the device. It is the application software for the device: the motor drive application, the valve application, or other application software. The user application communicates with the EtherNet/IP Adapter software layer by exchanging input and output assemblies and processing services requested over Explicit Messaging connections.

THE EVERYMAN'S GUIDE TO ETHERNET/IP

THE ETHERNET/IP SCANNER

WHAT'S INTERESTING?

What's interesting about EtherNet/IP Scanners is how few requirements there are for them in the EtherNet/IP specifications. Unlike EtherNet/IP Adapters, which must define an Object Model, support specific messaging and establish assemblies for cyclic data, the EtherNet/IP specifications are rather vague regarding EtherNet/IP Scanners. The EtherNet/IP specifications impose no requirements on how a scan list is created, how external devices can access that scan list, how a scanner should use an Electronic Data Sheet (EDS) or when a scanner should choose connected or unconnected messaging.

What many don't know about EtherNet/IP Scanners is that a scanner must support EtherNet/IP Adapter functionality, but that functionality does not require Implicit Messaging (as a target device). There is no requirement to support application objects, an Input Assembly object or an Output Assembly object. A scanner designer could choose to offer cyclic I/O Messaging, but the EtherNet/IP specifications do not require it. Most Programmable Controllers – the majority of scanner devices – choose to ignore it and don't offer any application objects or I/O Messaging.

EtherNet/IP Scanner devices must pass ODVA conformance testing to be considered compliant with EtherNet/IP specifications just like adapter devices.

WHAT'S IMPORTANT TO KNOW?

EtherNet/IP Scanners vary in their capabilities, both in terms of performance and features, but most exist to serve as the central controller of an EtherNet/IP network. Here's a summary of what you really need to know about EtherNet/IP Scanners:

1. An EtherNet/IP Scanner is the device on an EtherNet/IP network that sends outputs to EtherNet/IP Adapter devices and collects inputs from the EtherNet/IP Adapter devices.
2. EtherNet/IP Scanners are typically, but not always, programmable controllers. They collect inputs, execute logic and set new outputs just like all other programmable controllers, with the difference being that some subset of inputs and outputs is sent or received from the EtherNet/IP network.
3. EtherNet/IP Scanners can also be PC tools that make explicit connections to identify devices, configure devices, collect status information and monitor inputs and outputs on the network.
4. EtherNet/IP Scanners must support the required objects of the EtherNet/IP Object Model just like every other EtherNet/IP device. The Identify object, Message Router object, TCP object and Ethernet Link Interface must all be supported.
5. Scanners use the two message connections defined by the Common Industrial Protocol (CIP): Explicit Messaging connections and Implicit Messaging connections. Explicit Messages are service oriented and use request/response messaging to access the objects in other EtherNet/IP devices. Implicit messaging uses cyclic messaging to cyclically send outputs and collect inputs from EtherNet/IP Adapters.
6. Scanners use TCP (Transmission Control Protocol) transport for Explicit Messaging connections and UDP (User Datagram Protocol) transport for Implicit Messaging connections.
7. Although the EtherNet/IP specification defines other options for Implicit messaging, most scanner devices use cyclic data, via unicast I/O.
8. There is no standard way to configure an EtherNet/IP Scanner. Vendors of EtherNet/IP Scanner devices use a variety of ways to define the EtherNet/IP Adapter devices to manage and what inputs and outputs to access in those adapters.

WHAT ARE THE DETAILS?

At least one EtherNet/IP Scanner is required to form an EtherNet/IP network as EtherNet/IP Adapter devices can't initiate EtherNet/IP communications. An EtherNet/IP Scanner is the device that initiates communications with adapters, exchanges I/O data, reads attributes, writes attributes, requests services to be executed and monitors the status of connected devices. Most EtherNet/IP Scanners report adapter status – which adapters are present and which adapters are missing – to higher level applications.

In many ways, an EtherNet/IP Scanner is similar to an EtherNet/IP Adapter: both are CIP devices, support the same objects, have a nearly identical Object Model and use the same messaging interfaces (in opposite directions). In other ways, they are remarkably different. Unlike an adapter, a scanner needs a scan list of adapter devices to which it will connect and message. A scanner creates connections instead of waiting to accept connections offered to it. A scanner manages the connection process and chooses if communications with an adapter occur over a persistent interface (connected messaging) or a one-time, ad-hoc interface (unconnected messaging).

Scanner Object Model

An EtherNet/IP Object Model.[15] provides the external representation for the data the device designer wishes to present to the outside world. The Object Model of a CIP scanner device such as an EtherNet/IP Scanner or DeviceNet Master is composed of a set of objects with attributes that represent the data elements the device designer wants to make available over the network. Figure 22 illustrates the general structure of an EtherNet/IP Scanner Object Model. The two gray objects represent the objects that exist in every CIP device: the Message Router and the Identity object. The light gray objects with the dark borders represent objects specific to EtherNet/IP: the TCP Network object and the Ethernet Link object.

---

[15] An earlier chapter described CIP Object Modeling in great detail.

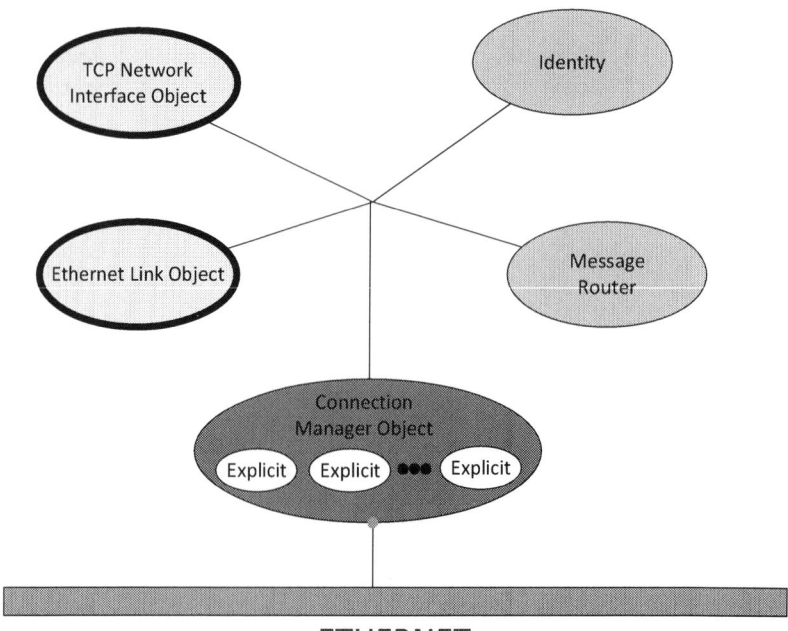

Figure 22 - EtherNet/IP Scanner Object Model

The dark grey Connection Manager object represents connections initiated by network tools and other scanners. These connections can come and go as needed and are mostly used to interrogate a scanners identity information. External connections, connections initiated by another scanner, are nearly always Explicit Messaging connections. Most EtherNet/IP Scanner devices don't accept implicit connections.

If you noticed that the model contains no representation for the connections with EtherNet/IP Adapter devices, you would be correct. Unlike the EtherNet/IP Adapter functionality where connections made with external devices are required to be represented by instances of the Connection Manager object, EtherNet/IP Scanners have no requirement to represent the connections they initiate to adapter devices. A scanner vendor could implement an object that represents those connections but would have to use a vendor-specific object.

Embedded EtherNet/IP Adapter: All scanner devices are required to support EtherNet/IP Adapter functionality, but that means less than it might initially appear. For example, there is no requirement for the adapter within a scanner to support any type of application object or support any Implicit Messaging. An embedded adapter could support implicit input or output connections with another scanner but few if any scanner device designers have chosen to implement functionality like that as it is not specifically required by the EtherNet/IP specifications. The Rockwell Automation programmable controller scanners, for example, don't accept implicit I/O connections.

In practice, the requirement to support adapter functionality simply means that an EtherNet/IP Scanner is required to accept, and process Explicit Messages as shown in Figure 22 above. Using that Explicit connection, a tool or another scanner could access any of the required objects (TCP, Ethernet, Identity, Router or Connection Manager) with service requests like Get Attribute or Set Attribute on any attribute in the scanner.

EtherNet/IP Scanner Configuration

All scanners require some sort of scan list. An EtherNet/IP Scanner connects to and messages each of the devices listed on the scan list.

The CIP and EtherNet/IP specifications do not specify how an EtherNet/IP Scanner obtains the scan list. Various vendors have designed mechanisms that fit their overall application objectives but no matter what mechanism is used, a scanner must obtain, at the minimum, the following data for each adapter in the scan list:

- TCP/IP Address
- Input Assembly instance ID and Size
- Output Assembly instance ID and Size

There are other possible data values that could be used to characterize the adapter like the Requested Packet Interval Rate or an optional Configuration Assembly Instance ID and Size, but the values listed above comprise the minimum set of data required to initiate communications with an adapter.

There are various common mechanisms that scanner devices sometimes use to create scan lists:

Hard Coded Scan List: It's relatively uncommon but sometimes the scan list is encoded in the software code of the scanner itself. Obviously, there is no possibility of adding or removing an EtherNet/IP device from the scan list without loading new applications software. That might seem like an extreme disadvantage but in a closed, proprietary environment where the scanner has fixed functionality that is unchangeable, encoding the scan list as part of the scanner software is practical.

Proprietary Scan List Creation: Different products support different mechanisms for interacting with users. Some have a phone or Linux/Windows application. Some have a keypad. Others have a front panel. There are many ways that devices interact with end users.

Some vendor devices extend their standard interface to include creation of the scan list. The advantage of proprietary scan list entry is that the scan list entry is now a seamless part of the standard configuration of the device. The disadvantage is that there are often data entry errors causing deployment delays and additional troubleshooting.

Connection Configuration object (CoCO): The Connection Configuration object (often shortened to "Coco") defines a common object-oriented interface to configure CIP connections within a scanner device. Tools like "RSNetWorks for EtherNet/IP" (from Rockwell Automation) can write the attributes of the CoCo object and set the scan list of a scanner device.

Electronic Data Sheet (EDS) Configuration: A more standard and more supported mechanism to configure a scan list is for a scanner to load the individual EDS files for all its devices on the scan list. From the EDS files, the scanner can programmatically identify all the information it needs to connect to and message each of the scanners. The advantage to this approach is that it eliminates any possibility of data entry errors. The disadvantage is that parsing an EDS file is difficult and time consuming. It is generally not something that low-end scanner devices support.

EtherNet/IP Scanner Connection Process: EtherNet/IP Scanners control all EtherNet/IP connections and initiate the connection establishment process. EtherNet/IP Adapters have no connections until an EtherNet/IP Scanner sends a connection initiation request. Scanners, by definition, uses unconnected messaging and the commands of the Unconnected Message Manager (UCMM) for connection establishment.

Unconnected Message Manager (UCMM) Explicit Messaging: The Unconnected Message Manager (UCMM) initiates Unconnected Explicit Message requests. This includes establishing both Explicit and Implicit Messaging connections with target devices. The Message Router object is the target of a UCMM request to establish an Explicit Message connection.

The Forward Open, Forward Close and Unconnected Send are the most important services that the UCMM uses for unconnected connection management and messaging. The Forward Open contains important connection related parameters, many of which are used to configure the instance created by the Connection Manager object to manage the new connection. The service data included with the Forward Open service request includes items particular to the connection type (Explicit or Implicit). For implicit connections, items include the RPIs (Requested Packet Intervals), connection paths and transport types. The Forward Close is the command used to terminate a connection and the Unconnected Send is the command used to route Explicit Messages without opening a connection.

Many scanner devices choose to always use UCMM for unconnected Explicit Messaging, rather than supporting Explicit Messaging via a connection to limit resource requirements. Network tools are examples of devices that always use unconnected Explicit Messaging services.

EtherNet/IP Message Connection Process

The process to create a connection between an EtherNet/IP Scanner and an EtherNet/IP Adapter varies slightly depending on the type of connection you are creating. There are two ways of messaging a CIP device; connected messaging and unconnected messaging. In connected messaging a connection is created, and messages are exchanged over that connection for some period of time. In unconnected messaging, a single message and its response is exchanged, and the connection is closed. Both are common. Connected messaging is used for I/O connections where inputs and outputs will be exchanged until the next power cycle on the machine. Unconnected messaging is used by both programmable controllers and network tools to identify devices, get and set device configuration and monitor device operation.

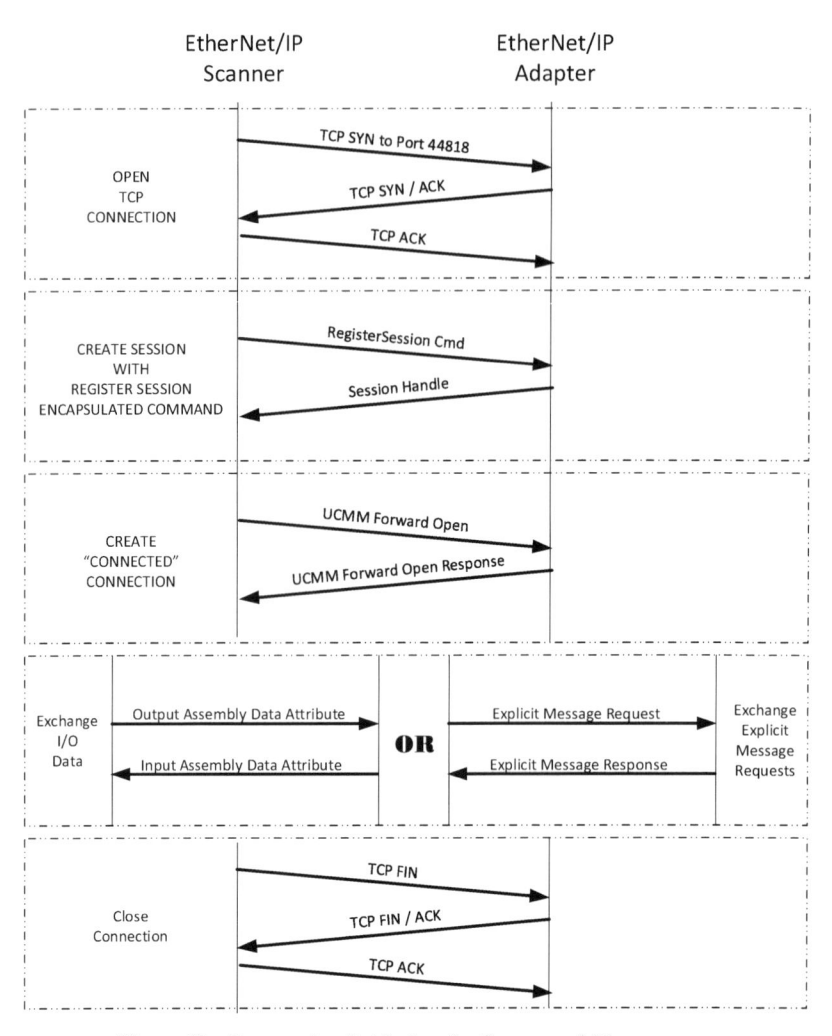

Figure 23 - Connection Initiation for Connected Messages

Figure 23 illustrates the process used to establish a connection and exchange either Implicit or Explicit Messages over that connection. There are three steps required prior to exchanging data over a connected message interface:

1. **Establish a TCP connection:** A TCP connection is required to configure the connected connection that will be used to exchange the explicit or Implicit Messages. A TCP connection is created using a three step process where the scanner sends a TCP SYN to the EtherNet/IP port on the adapter (Port #44818). The adapter acknowledges that with a TCP SYN/ACK. Lastly the TCP ACKs that message and the TCP connection is established.

2. **Register an Encapsulation Session:** All EtherNet/IP messages are encapsulated using the Encapsulation Protocol. The second step in the connection sequence is to establish a session where the encapsulated command can be exchanged. In this step, a RegisterSession encapsulation command is transmitted to the adapter. The adapter responds with a RegisterSession response and a handle to identify the Session.
3. **Create the desired connection:** In this step an open connection request, the ForwardOpen command is encapsulated in the Encapsulation Protocol. The adapter then creates the explicit or implicit connection as described by the ForwardOpen

Once these steps are complete, the connection is ready and Explicit or Implicit Messages can be exchanged over the connection. Once, no longer needed the EtherNet/IP Scanner can close the TCP connection using the TCP FIN command. That command releases all the resources allocated by the TCP connection and any outstanding encapsulation commands or CIP explicit connections. CIP implicit connections use UDP messages, so they are unchanged when the TCP connection is closed.

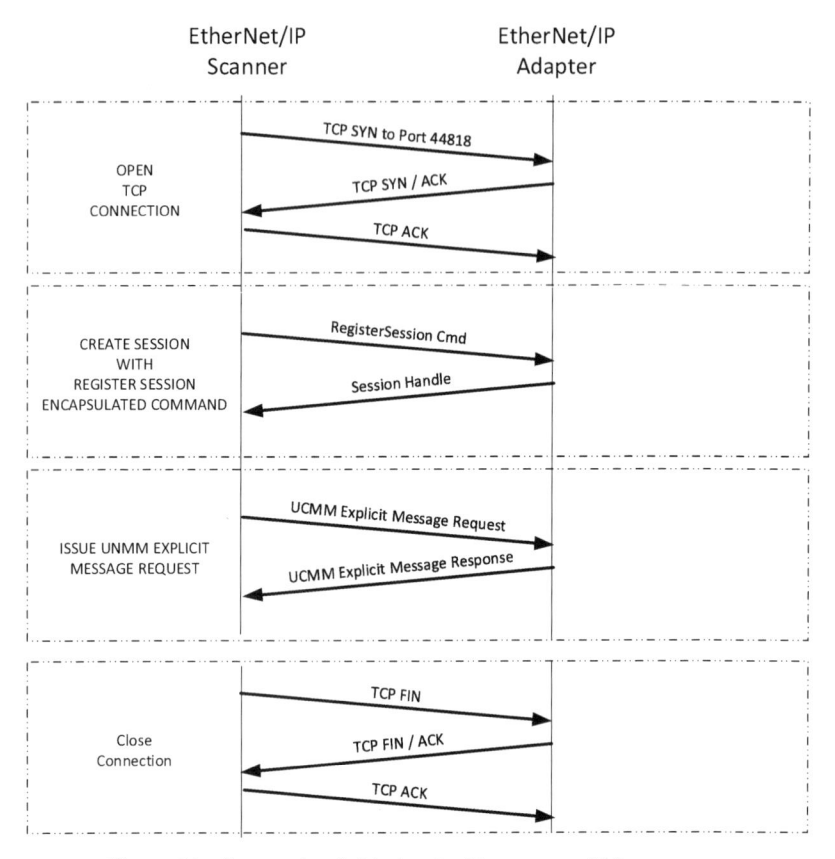

```
                EtherNet/IP              EtherNet/IP
                 Scanner                   Adapter

                         TCP SYN to Port 44818
           OPEN    ─────────────────────────────►
           TCP              TCP SYN / ACK
        CONNECTION ◄─────────────────────────────
                              TCP ACK
                   ─────────────────────────────►

                          RegisterSession Cmd
         CREATE SESSION ──────────────────────►
            WITH
         REGISTER SESSION    Session Handle
       ENCAPSULATED COMMAND ◄──────────────────

                         UCMM Explicit Message Request
       ISSUE UNMM EXPLICIT ──────────────────────────►
       MESSAGE REQUEST   UCMM Explicit Message Response
                         ◄──────────────────────────

                              TCP FIN
                      ──────────────────────►
           Close             TCP FIN / ACK
         Connection   ◄──────────────────────
                              TCP ACK
                      ──────────────────────►
```

Figure 24 - Connection Initiation for Unconnected Messages

Figure 24 illustrates the process used to establish and exchange a message over an unconnected message interface. Note that the process is nearly identical to the connected message interface. Instead of connections being created by the Encapsulation Protocol, the UCMM sends an Explicit Message request. Once that is complete, the connection is terminated.

EtherNet/IP Scanner Messaging

CIP defines two kinds of connections; Explicit Message connections and Implicit Message connections. Explicit Message connections are used to open and close connections, read and write attributes of the Object Model and request services to be performed. Implicit message connections are used to transfer inputs from adapters to a scanner and outputs from the scanner to adapters.

Explicit Message Support: Scanners use Explicit Message connections to create connections with other EtherNet/IP Scanners and Adapters.[16], access the Object Model of other EtherNet/IP devices and provide their internal Object Model data to other EtherNet/IP Scanners and network tools. Scanners can generally support any number of Explicit Message connections with the number of connections typically limited by available resources.

Incoming Explicit Message connections, connections initiated by other scanners, are used by network tools and other scanners to access the scanner's Object Model. A network tool, for example, may open a connection to read the attributes of the scanner's Identity object. Another scanner, like a programmable controller, might use an Explicit Message connection to exchange data by reading or writing a vendor-specific object in the scanner.[17]

Outgoing Explicit Message connections, connections initiated internally, are used to access the Object Model of other EtherNet/IP devices. In some applications, a scanner might use Explicit Message connection to exchange I/O data with an adapter on an intermittent basis instead of using a cyclic connection. Adapter input data can be accessed over an Explicit Message connection by reading the data attribute of the adapter's Input Assembly or attributes of application objects. An adapter's output data can be accessed over an Explicit Message connection by writing the data attribute of the adapter's Output Assembly or the attributes of application objects.

Explicit Messages are sometimes used to transfer I/O when the inputs or outputs are required on an as-needed basis. For example, if the EtherNet/IP Adapter was a temperature probe, the application engineer might choose to only read the temperatures when the data is required using Explicit Messaging instead of using a cyclic connection and wasting bandwidth. While this is possible, it would be unusual and is very unlikely to be used in applications where the scanner is a programmable controller. In a programmable controller, logic is required to trigger an Explicit Message and many programmers dislike writing extra logic.

Unlike an EtherNet/IP Adapter, only incoming Explicit Message connections are managed by an instance of the Connection Manager object. The EtherNet/IP Scanner does **not** require outgoing Explicit Messages, connections initiated internally, to be managed by the Connection Manager Object. Scanner developers are free to implement, control and monitor outgoing Explicit Message connections in any way they choose.

---

[16] Using an UnConnected Message Manager (UCMM) connection
[17] That would be a completely proprietary connection.

Implicit Message Support: Scanners use Implicit Message connections to transfer inputs from an adapter and to send outputs to an adapter. A scanner can have input and output message connections with as many adapters as its resources permit. Scanners generally support one input connection and one output connection with each adapter but there are adapters that are either input-only or output-only.

Unlike an EtherNet/IP Adapter, only incoming Implicit Message connections, connections initiated by another scanner, are managed by an instance of the Connection Manager object. The EtherNet/IP Scanner does **not** require outgoing Implicit Messages – connections initiated internally – to be managed by the Connection Manager Object. Scanner developers are free to implement, control and monitor their own outgoing Implicit Message connections in any way they choose.

Most connections to adapters are unicast – one scanner to one adapter – but multicast operation is possible. IGMP (Internet Group Management Protocol) is used for multicast operations. In multicast, the Ethernet switch manages subscribers and publishes the Implicit Messages to all the subscribers on the multicast list. Multicast operation is supported by many Ethernet switches. Scanner devices must subscribe to multicast messaging on a per-adapter basis when the Implicit connection is opened. Scanner devices must unsubscribe from multicast messages when the multicast address is no longer needed (connection timed out).

A View Under the Hood

If you want to understand how an EtherNet/IP Scanner is architected, it's not all that complicated. Figure 25 shows the various software layers you might find in an EtherNet/IP Scanner.

THE EVERYMAN'S GUIDE TO ETHERNET/IP

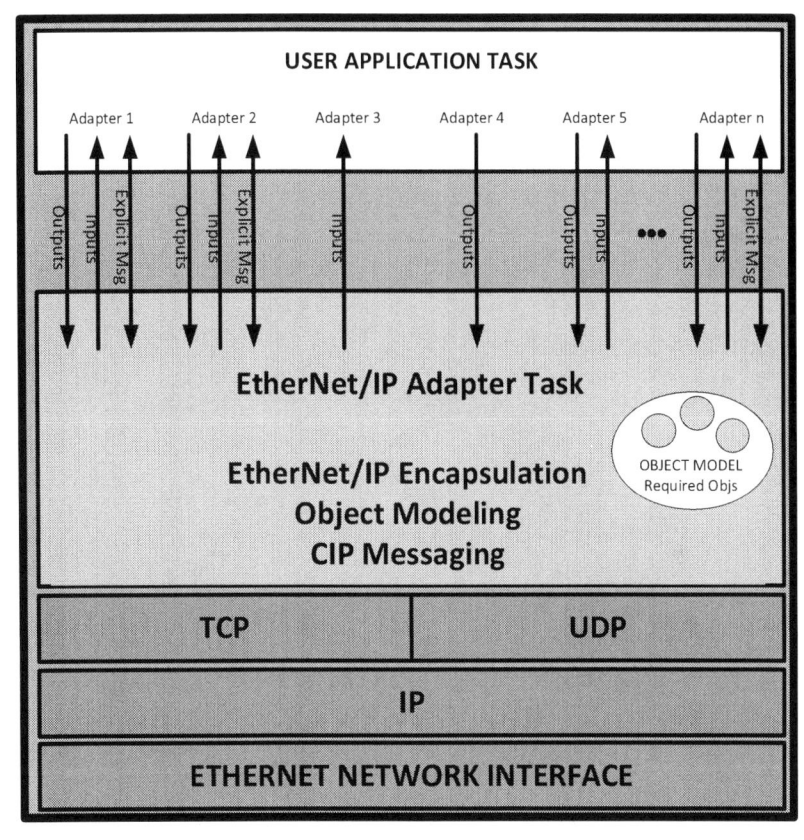

Figure 25 – Ethernet/IP Scanner Software Structure

There are three main layers in this architecture:

1. **The TCP/IP Stack:** the TCP/IP stack (dark gray area) provides the communication protocols needed to move messages across an Ethernet network. These protocols include (but are not limited to) TCP (Transmission Control Protocol), UDP (User Datagram Protocol), IP (Internet Protocol) and the Ethernet physical interface. The TCP and UDP layers provide transports for moving messages between an EtherNet/IP Adapter and an EtherNet/IP Scanner. The IP and Ethernet Physical Interface provide low-level routing of messages between the nodes of the Ethernet network.

2. **The EtherNet/IP Scanner Software:** the EtherNet/IP Scanner software layer (gray areas) provides both CIP (the Common Industrial Protocol) and the EtherNet/IP encapsulation layer.[18] This layer is where CIP Explicit and Implicit Messages are formed

---

[18] EtherNet/IP Encapsulation is discussed in an earlier chapter in this section.

and encapsulated in the Encapsulation Protocol. The EtherNet/IP Scanner sends Forward Open messages to establish the Explicit and Implicit connections requested by the user application. This layer also schedules the cyclic input and output messages that are exchanged with the individual adapter devices at the Requested Packet Interval (RPI) chosen by the scanner.

3. **The User Application:** the user application is the software that implements the basic functionality of the device including any device configuration, device operation and processing of the adapter inputs and creation of the adapter outputs. For many programmable controllers, the user application interacts with ladder logic or a similar logic engine to provide meaningful output data and process the input data. Many programmable controllers have a separate communication engine or processor to handle the connection maintenance and data handling required to work with many adapter devices.

You'll notice from Figure 25 that the application layer has various kinds of connections with adapter devices. In some cases, it has explicit, implicit input and implicit output connections. In other cases, it may just have an explicit connection, an input only connection or an output-only connection.

COMPLIANCE TESTING

Specifications, unfortunately, are sometimes written in a way that is subject to interpretation. In the history of software development, it's been proven time after time that it is very difficult to write a consistent, clear, easily understood technology specification. Humans are fallible – both the humans writing the specifications and those implementing what they read. The more complex the specification, the worse this problem can be.

Specifications like the one for CIP and the extension to it for EtherNet/IP are long, complicated and sometimes difficult to read and interpret. When you have a very lengthy and complex specification (it would be better to drop a brick on your big toe than the heavy EtherNet/IP specification) and hundreds of programmers all over the world developing products with it, there are going to be misinterpretations and inconsistencies among implementations. And unfortunately, if those inconsistencies aren't resolved, they are going to surface at a manufacturing facility where some Control Engineer is just trying to get a machine to run.

The ODVA answers this problem with a very well-engineered and well-run Compliance Test Process that is required for all products using EtherNet/IP.

Benefits of Compliance Test Process

The ODVA is the legal owner of the Common Industrial Protocol (CIP) and its derivative technologies: EtherNet/IP, ControlNet and DeviceNet. It provides a non-exclusive license to those technologies to product vendors using those technologies. The license confers the legal right to use those technologies and associated logos and trademarks but also requires vendors to fulfill certain responsibilities.

The most important vendor responsibility is to certify all products based on those technologies through the Compliance Test Process.

To ensure consistency and interoperability among licensed products using these technologies, the ODVA has implemented a very well-run and well-defined test system. The Compliance Test Process serves both vendors and end-users on the plant floor. Vendors with certified products vastly decrease future support costs by ensuring that their product is compliant with the technical specifications. End users can choose products with confidence, knowing that they there will not be major interoperability issues among products certified by the Compliance Test Process.

ODVA promises vendors five benefits from its conformance test process:

- Efficiency – The conformance test is done in an environment with a diverse range of products in a homologated test environment from many different vendors with a range of approaches to product implementation. That is an environment that few vendors can afford to duplicate.

- Know-How – The testware that is used for the certification test is developed in close cooperation with the technical committees that maintain the EtherNet/IP technical specification. The contributors to the specifications and testware have a critical mass of know-how in ODVA technologies, standards and compliance that no vendor can duplicate.

- Objectivity – The ODVA operates its test labs independently from any member or vendor providing added assurance that ODVA's determination of compliance is made without bias and administered equally regardless of the vendor submitting the product(s) for evaluation.

- Confidentiality – All interactions with ODVA's technical staff are completely confidential. Not until final certification is the fact of a product's acceptance made public through an announcement on the ODVA website.

- Better Released Product – The compliance test process and related discussions with ODVA technical staff assists vendors in creating much better products with less potential to cause issues on the manufacturing floor.

The Term of Usage Agreement

Over the last ten years, open source software has grown very popular. With open source, source code for operating systems, products, technologies and many other things are made freely available to anyone. It's very attractive to new entrepreneurs as there are no fees and no ongoing license costs. EtherNet/IP is not open source technology. The ODVA and the EtherNet/IP development toolkit providers have not posted any open source for any CIP technology and are unlikely to do so in the future. There are some websites that have some open source EtherNet/IP source code. Even though product vendors may use that source code, it is perilous as that source code is unlikely to be maintained as well as source code from one of the EtherNet/IP development kit providers.

No matter how a product vendor procures its EtherNet/IP source code, they must still register with the ODVA to legally use CIP and EtherNet/IP technology as CIP and EtherNet/IP are technologies wholly owned and controlled by the ODVA. The ODVA maintains the technology, promotes it, supports it and organizes technical activities to extend and maintain it. None of the ODVA licensed technologies can be legally used without registering with the ODVA.

The first step in registering as an EtherNet/IP product vendor with the ODVA is to sign the ODVA Terms of Usage agreement (Figure 26). The Terms of Usage agreement is a license agreement that allows companies actively developing EtherNet/IP products to legally sell EtherNet/IP products and use the ODVA logos.

Figure 26 – ODVA Terms of Usage Agreement (1st Page)

The Terms of Usage spells out the specific obligations and responsibilities that a vendor has when using ODVA licensed technology. The agreement is long and lawyerly but in general, the vendor is obligated to purchase and maintain a specification, test products in an ODVA certified test facility, submit products for testing upon demand and grant ODVA legally authority over patents regarding the functioning of EtherNet/IP in their product.

The Declaration of Conformance

The purpose of the Conformance Test is to verify that the licensed product implementation is compliant with the CIP and EtherNet/IP specifications. It begins with the product vendor submitting a Test Service Order Form on the ODVA website. The Test Service Order process requires the vendor to declare:

- The CIP derived technology: EtherNet/IP, ControlNet, DeviceNet
- The product class: Originator (scanner) or Target (adapter)
- The Test Service provider location: US, Europe, Asia
- The Product Identity information: Vendor ID, Product code, Name, Revision

Once the Test Service Order is accepted, the product goes into the queue and a test date is assigned. Prior to the date of test, the vendor must submit a representative sample of the product and some additional technical documentation. Product vendors are sometimes confused as to what constitutes a "representative sample" for large, complicated devices like a chiller or a robot. There is no requirement for actual live data to be transferred to the test controller. All that is required is for the unit to respond to EtherNet/IP messages so in many cases only the electronics control module for the device under test needs to be provided.

Sometimes product vendors are developing a series of closely related products using very similar hardware and software. A valve family with 8, 16 and 32 valve units is an example of a closely related product family. In these cases, the product vendor can request family certification. Only two representative products from the product family are required to certify a product family.

On the date of the test, access to the test facility is limited to test service personnel. Everything related to testing – the vendor, the product(s) being tested and all results – are kept strictly confidential. Optionally, the product vendor can send a representative to observe the test and, if needed, provide immediate product enhancements, but in practice, few vendors attend the test.

If the product cannot achieve a passing result, the experts at the Test Service facility are available to decipher the test results and consult on reasons for the failure. That consultation is, however, limited. Additional consulting can be purchased from the ODVA if a product vendor needs more extensive consultations on their product implementation.

If successful, the ODVA issues a Declaration of Conformance (DOC). The DOC is the ODVA's statement that a product (or product family) is compliant with the CIP and EtherNet/IP specifications. The vendor is then authorized to affix the ODVA CONFORMANT logo mark (Figure 29) on the product that has received the DOC. The logo mark is required to be placed on the informational Product label or some similar place on the product.

Note that there is an ODVA Brand Standard and Identity Guideline[19] that describes in great detail exactly how the terms and logos may be used. For example, reproduction of the ODVA Conformant trademark in word form is generally not allowed. Check the Brand Standard and Identity Guide for if and when the words "ODVA CONFORMANT™" might be acceptable.

The ODVA is required by its bylaws to post every Declaration of Conformance on its website. The product name, vendor and other information regarding every conforming product are listed on its website.

ODVA Branding Guidelines and Logos

ODVA's Brand Standards+Identity Guideline is very particular regarding the use of trademarks and logos. Maintaining brand equity is important, and ODVA members are required to use all trademarks and logos as directed. Some of the more prominent rules are listed here, but you are encouraged to become familiar with the *Guideline* as it is the final authority on ODVA trademarks and logos.

1. Members are encouraged to display the relevant technology logo in a visually prominent place on the product. It is not acceptable to use the word mark form of the logo except in cases where there is no room for the logo in its minimum size configuration.
2. Icons for CIP Services, CIP Safety, CIP Energy, CIP Sync, CIP Motion and/or CIP Security, if certified for the product by the ODVA, can be placed on the product but must be listed in the order shown on the DOC.
3. ODVA trademarks and word marks can be used on product collateral but they must match the acceptable formats as shown in the Brand Standard+Identiy Guideline.
4. Use wording such as "This product is designed in accordance to The EtherNet/IP™ Specification" and not "This product is ODVA CONFORMANT."
5. Members are encouraged to use the "ODVA MEMBER" logo alone to indicate their affiliation with ODVA.

Table 11 – Five Important ODVA Branding Guidelines

---

[19] See the ODVA website (https://www.odva.org/)

THE EVERYMAN'S GUIDE TO ETHERNET/IP

ODVA Licensed Logos

EtherNet/IP®

Figure 27 – Ethernet/IP Technology Only Logo

EtherNet/IP®
ODVA

Figure 28 – Ethernet/IP And ODVA Combined Logo

ODVA
CONFORMANT

Figure 29 – Logo For Conformant Products

ODVA
MEMBER

Figure 30 – ODVA Member Logo

JOHN S. RINALDI

ETHERNET/IP FOR END USERS

If you're an end user of EtherNet/IP, and many of you are, once you have found an EtherNet/IP Adapter that meets your physical and functional requirements, you should consider three questions before selecting that device:

1. Did the device pass certification at the ODVA lab? (Declaration of Conformity)
2. How easy will it be to integrate it into my controller? (Device Integration)
3. How easy is it to use the device's data? (Data Integration)

DECLARATION OF CONFORMITY (DOC)

If a device you plan to use passed certification, the device vendor will make available the Declaration of Conformance (DOC) for the device and use the ODVA Conformant logo (Figure 31) in the product literature. You will want to check the versions of the product that received the DOC against the current revision that you are using. There are vendors that received their DOC years ago, and though they've continually rolled out their product, they haven't tested it with the latest conformance test in many years.

Figure 31 – ODVA Conformance Logo

THE EVERYMAN'S GUIDE TO ETHERNET/IP

DEVICE INTEGRATION

Once a new device arrives onsite, you'll need to get it integrated with your network and your controller to get it deployed. There are at least two generic steps to that process beyond the physical mounting and electrical integration into the network.

The first step is to configure the device as an Ethernet node. Unfortunately, this is a very proprietary process that varies from vendor to vendor. Somehow the device will need a TCP/IP address, subnet mask and default gateway address. There are endless ways that you can set these values: DHCP, a front panel, a utility program, dip switches, your cell phone and many more.

Once your new EtherNet/IP device is properly configured, you can now integrate it with your EtherNet/IP Scanner. Most EtherNet/IP Scanner devices are Rockwell Logix brand controllers, so we'll discuss that integration here.

There are five possible mechanisms for integrating your device with a Rockwell Automation Logix controller. Each of these will be described here, from the least useful to the most highly integrated.

1. **You have a minimum functionality EDS (EDS "stub" File)**

All EtherNet/IP devices must make available an Electronic Data Sheet (EDS) file, but there is no requirement for that EDS file to contain much more than identity information. If you have a device with a minimum EDS, you can add that device to the PLC device tree as a generic device. As a generic device, you manually input the number of bytes of input and the number of bytes of output.

There is no advantage to using devices with minimum EDS files and one big disadvantage. The big disadvantage is that you'll have to use PLC logic to decode the Input Assembly received from the device and to build the Output Assembly to send to the device. The program logic won't automatically know how to interpret any of the bytes received from or sent to the device. If the device is simple, like an 8-channel valve, that might be easy. If it's a more complicated device with a hundred bytes, it will be a nightmare.

2. **Simple EDS (Rockwell Automation Calls this "Non-Licensed AOP EDS")**

Some vendors provide an EDS with more than a stub file. The EDS describes the configuration data, assemblies and other information in the EDS. These files are quite useful to the Rockwell Automation programming tool:

They allow you to choose from a predefined list of connections (combinations of input, output and configuration assemblies).

They expose generic configuration options like RPI and Multicast/Unicast.

They allow instantiation of multiple connections at once (note that you have to manage the data consistency of those multiple connections).

3. **Use an Add On Instruction (AOI)**

    Some vendors provide an Add-on instruction (AOI) with their device and EDS. AOIs are an interesting feature of Rockwell Automation PLCs. Add-On Instructions are user defined instructions to capture some set of instructions (or an algorithm) and use it as one instruction. Some vendors create an AOI that knows how to decode a device's Input Assembly or encode an Output Assembly. It's usually not a perfect solution for integrating EtherNet/IP Adapters but not a bad one either.

4. **EDS with Licensed Key (Rockwell's "Licensed EDS AOP")**[20]

    Vendors that have worked with Rockwell to get a RA Licensed AOP can provide an EDS that is highly integrated with the programmable controller tool set. These devices can use the programmable control configuration wizard to configure a device as well as download one-time configuration to a new device.

5. **PLATINUM Option (Rockwell's Custom AOP)**

    Vendors with very close ties to Rockwell can get special integration of their devices into the Rockwell programmable controller tool set. This makes their adapters appear alongside Rockwell adapters in the device tree where EtherNet/IP Adapter devices are selected.

DATA INTEGRATION

Data integration should not be a concern, but many EtherNet/IP Adapters have been created that don't work well with Rockwell Automation programmable controllers. Some devices use ASCII strings that are too long for the string data types in a PLC. Some use assemblies with boundary errors – misalignment of floating point values for example. Others use data types that aren't supported by programmable controllers.

Over the years, vendors have improved so that these problems are now relatively uncommon.

---

[20] See the Chapter on EDS Files for more information on the Add-On-Profile

ETHERNET/IP FOR SW ENGINEERS

If you want to understand how an EtherNet/IP Adapter is architected, it's not all that complicated. It's just that there are multiple ways of looking at it.

PROTOCOL SUITE ARCHITECTURE

One way is to compare the network layers in an adapter to the network layers specified in the OSI (Open Systems Interconnect) Model (Figure 32). If you do that, you'll immediately notice that an EtherNet/IP Adapter doesn't correspond directly to the architecture of the OSI software model. There are no specific session, presentation and application layers in the EtherNet/IP model. Instead, various parts of CIP fill those roles.

OSI MODEL ETHERNET/IP MODEL

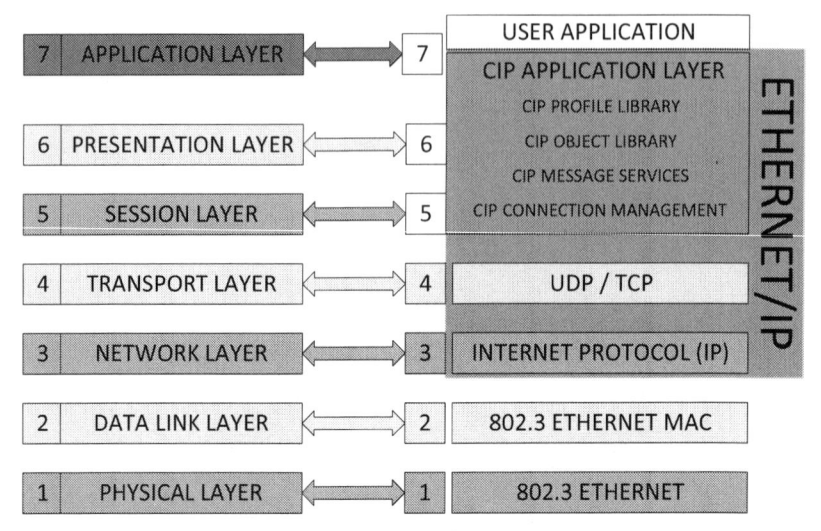

OSI		Ethernet/IP
7	APPLICATION LAYER	7 — USER APPLICATION / CIP APPLICATION LAYER (CIP PROFILE LIBRARY)
6	PRESENTATION LAYER	6 — CIP OBJECT LIBRARY / CIP MESSAGE SERVICES
5	SESSION LAYER	5 — CIP CONNECTION MANAGEMENT
4	TRANSPORT LAYER	4 — UDP / TCP
3	NETWORK LAYER	3 — INTERNET PROTOCOL (IP)
2	DATA LINK LAYER	2 — 802.3 ETHERNET MAC
1	PHYSICAL LAYER	1 — 802.3 ETHERNET

Figure 32 – OSI vs. Ethernet/IP Software Model

What you will learn from this comparison is that the access to CIP functionality over TCP, UCP, IP and Ethernet is what defines EtherNet/IP. Just as accessing CIP functionality over CAN is what defines DeviceNet and accessing CIP functionality over the ControlNet Physical Layer is what defines ControlNet.

TCP/IP PROTOCOL SUITE

As a software engineer, you'll immediately note from Figure 32 is that the TCP/IP protocol suite is integral to the building of an EtherNet/IP device. How your application interacts with other EtherNet/IP devices and other non-EtherNet/IP devices on the network is dependent on the TCP/IP protocol suite you choose for your product.

The TCP/IP protocol suite is the link between every single one of the Ethernet applications supported on your device. It's integral to supporting IGMP for multicast support, PING for network access, web page access (HTTP), file transfer (FTP) and every other component of network functionality you might have on your device. It provides all the know-how to make connections with other devices, route messages through an Ethernet network, packetize large streams of data and much more. The TCP/IP Stack is critical software not only to support your web server but also for EtherNet/IP, PROFINET IO and Modbus TCP.

THE EVERYMAN'S GUIDE TO ETHERNET/IP

One of the key features of TCP/IP is that is isolates you from the low-level interaction with your Ethernet hardware: your Ethernet PHY (Physical Interface) and Ethernet MAC (Media Access). The TCP/IP Stack has to understand exactly what kind of hardware you have in your computer and how to transfer a message back and forth to that hardware. At the lowest levels, a TCP/IP Stack builds a message and must then send it bit by bit to the Ethernet MAC (Media Access Controller) after which it will be converted to electrical signals by a PHY (Physical Interface). This means that the TCP/IP Stack must not only work with CIP to encode and transfer encapsulated messages, but it must also be intimately familiar with your hardware.

For that reason, you can't just pick up one TCP/IP Stack and reuse it for another device. It has to have the right drivers for that device. It must know and understand exactly what PHY and MAC your device is using.

There are TCP/IP protocol suites that come with operating systems: Windows and Linux both include TCP/IP. There are silicon vendors that provide a TCP/IP stack at no charge as an incentive to use their hardware. And there are vendors that are in the business of providing TCP/IP protocol suites.

But the problem for software vendors in manufacturing automation is that sometimes these protocol suites don't work quite right. Or they don't work properly to pass the Conformance test for EtherNet/IP or the Conformance test suite for PROFINET IO.

To assist device vendors in selecting a TCP/IP stack that will pass the EtherNet/IP Conformance test, the ODVA has published a set of minimum requirements for a TCP/IP protocol suite. You can get this from the ODVA, but the requirements listed in that document are pretty basic. The TCP/IP protocol suite must support the:

- Internet Protocol (IP version 4) (RFC 791)
- User Datagram Protocol (UDP) (RFC 768)
- Transmission Control Protocol (TCP) (RFC 793)
- Address Resolution Protocol (ARP) (RFC 826)
- Internet Control Messaging Protocol (ICMP) (RFC 792)
- Internet Group Management Protocol (IGMP) (RFC 1112 & 2236)
- IEEE 802.3 (Ethernet) as defined in RFC 894

The ODVA specification requires that a device support IP, TCP, UDP, ARP and other protocols. The RFC (Request For Comments) is the IEEE's designation for an internet specification. For example, RFC 826 describes exactly how the ARP (Address Resolution Protocol) must function.

OBJECT MODEL ARCHITECTURE

Another view of the EtherNet/IP Adapter architecture is what you find in Figure 33. In this view, the architecture of an EtherNet/IP Adapter is shown as the objects it supports. The adapter is composed of the set of required objects that must exist in every EtherNet/IP device and a set of Application objects, which provide the interface to the real world.

That would be correct, but it misses the point because this representation lacks the most important component of an EtherNet/IP Adapter: the connection to the physical world. It is a more informative view of an EtherNet/IP Adapter, but it hides the network layers that make a CIP device an EtherNet/IP device. That functionality is buried in the Connection object that manages those CIP Message Connections. What is shown in this Object Model view are the Application objects that form the physical interface of the EtherNet/IP Motor Drive, Valve Block or other device.

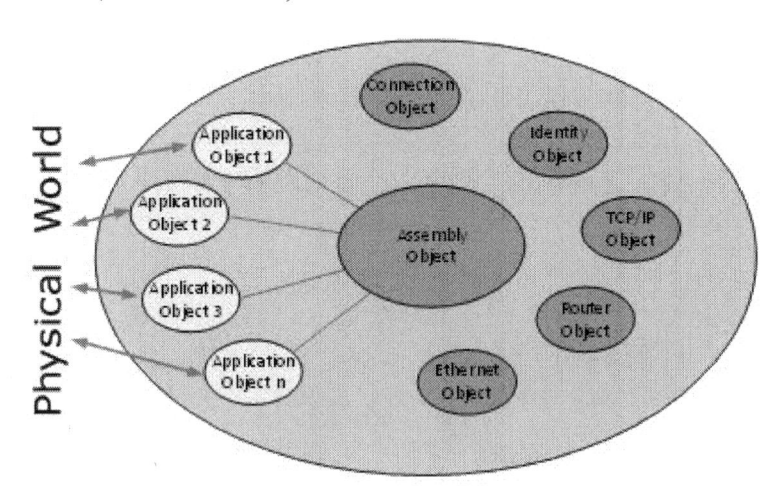

Figure 33 – EtherNet/IP Object Model Physical Connections

This view implies, but doesn't make explicit, the separation of the device application from the EtherNet/IP implementation.

TASK ARCHITECTURE

Another way to look at the architecture of an EtherNet/IP

Adapter is presented in Figure 34. This view presents a task-oriented view of how a user software application might interface to an EtherNet/IP Adapter and CIP communications software. In this diagram, the user application task is the software that does the actual work of controlling the linear actuator, running the drive motor or turning the pneumatic valves on and off.

In this view, the user application task interfaces with EtherNet/IP Adapter software using Explicit and Implicit Message connections. The application task passes real world inputs (the Input Assembly instance data attribute) to the EtherNet/IP Adapter task, which encodes it as an encapsulated CIP message and sends it on the cyclic input channel to the originator of the connection. For outputs, the originator of the connection passes outputs as encapsulated CIP messages to the EtherNet/IP application task, which decodes the encapsulation and CIP messaging and passes the outputs (the Output Assembly instance data) to the user application task.

Explicit Messages are handled much the same way. Decoded Explicit Messages containing service codes to execute are also passed to the application task for processing. Explicit responses are returned to the EtherNet/IP Adapter task for encoding as EtherNet/IP messages to be sent to the originator.

It's important to note here that the user application program knows nothing of the communication mechanism. It simply processes data it receives that is defined by the Output Assembly instance. It collects data, formats it as Input Assembly instance data and sends that to the communications task. The user application task is totally unaware over what network and how that data was encoded, so many of these user application tasks could be used with other networks like PROFINET IO and Modbus TCP.

JOHN S. RINALDI

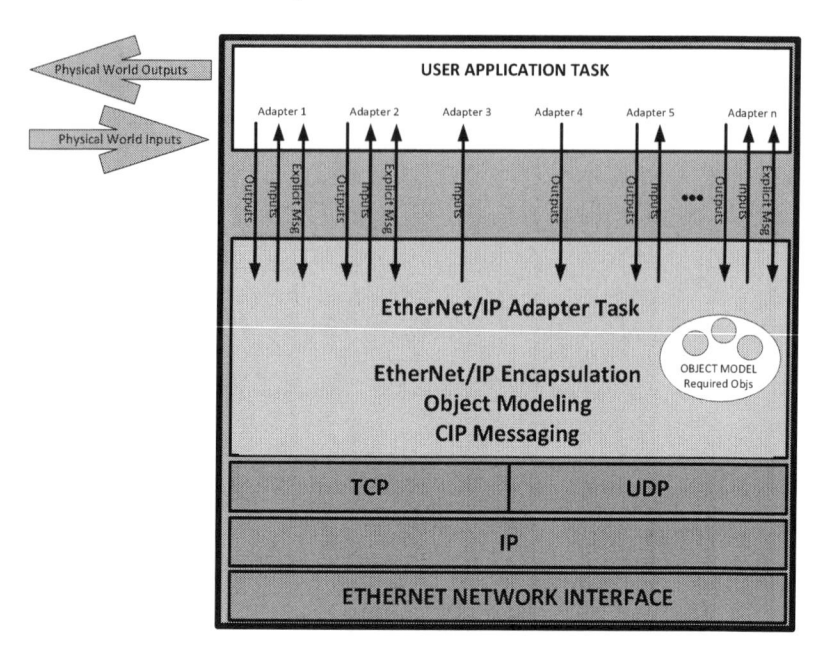

Figure 34 - User Application Interface

This view is, of course, a simplified software structure and doesn't provide the exact software structure for an EtherNet/IP Adapter, but it is, nevertheless, instructive for understanding the architecture of a CIP device.

THE EVERYMAN'S GUIDE TO ETHERNET/IP

EDS FILES & CONFIGURATION

An Electronic Data Sheet (EDS) is simply an ASCII file that describes how a device can be used on an EtherNet/IP network. The concept excited customers back when EtherNet/IP was new but, while important, it hasn't become as useful as most imagined back when it was introduced. The truth was that until a few years ago, an EDS file wasn't necessary for the Conformance Test Lab, much less used in the field. Now that programmable controllers make more use of EDS files, they have become very important and a key component of an EtherNet/IP Adapter device.

Only a minimal amount of information is required of an EDS file, and the amount of information stored in an EDS file varies from device to device. Some manufacturers store only that minimum, required, amount of information, sometimes nothing more than just device identification. Others store the details of every object and every attribute in the EDS including valid data ranges and attribute data types.

At the minimum, an EDS file conveys the device information required for a EtherNet/IP network tool to recognize the device. But most device manufacturers now realize that it is important for the EDS file to describe the contents of the Implicit Messages exchanged between an EtherNet/IP Scanner and an EtherNet/IP Adapter. Scanners also have EDS files, but the only use for a scanner EDS is to provide identity information.

When EtherNet/IP was first introduced, EDS files were often shipped with a device in some media format like a CD. Now EDS devices are available on the device manufacturer's website. Some devices with extended data storage contain the EDS file internally within the device.

Users often think that EDS (Electronic Data Sheet) files provide more than they really do. EDS files are ASCII files that describe an adapter device to the network. At a minimum, the EDS files describe what the device is, its model number, revision and most importantly, the size of the I/O Assemblies. It can also describe all the objects in the adapter, what attributes exist, if they are readable or writeable and sometimes if the vendor wants to be complete, the range of values for each of these attributes.

TYPES OF ADAPTERS AND EDS FILES

EtherNet/IP Adapter devices fall into a couple of different categories depending on the level of sophistication of their EDS files.[21]:

- **Adapters with no EDS File** – These devices typically come from 3rd party vendors with no relationship to Rockwell Automation. The EDS may not exist or not be available. There isn't much a Logix programming tool can do with these devices. It can find the device on the network and automatically capture the TCP/IP Address but that's all. The device must be manually added to the EtherNet/IP device tree and the I/O Assembly sizes must be entered manually. The I/O Assemblies must be manually assigned to some area of the programmable controller's memory (Ints, Dints, Floats…etc.). And the Configuration data, the block of data that a ControlLogix PLC sends to an adapter with the EtherNet/IP Forward Open, must be entered manually.

- **3rd Party Adapters with an EDS File** – Rockwell calls these devices EDS AOP devices where AOP means Add On Profile (See the discussion below on AOPs). Besides populating an IP Address, the EDS now is used to automatically populate the I/O Assembly Sizes. Configuration data is still entered manually and there is no way to access the I/O Assemblies other than as an array of raw data.

- **Adapters from AOP Partner EDS Files** – Rockwell provides special functionality for devices from vendors accepted into its AOP program. They call this functionality Add On Profiles (AOPs as opposed to EDS AOPs). AOPs allow RSLogix to not only find and load the IP Address and Assembly Sizes but also to define the Assemblies as a series of Tags for ease of use in the Controller logic. Configuration data can be defined for loading

---

[21] The vast majority of EtherNet/IP Adapters are used with Rockwell Automation Logix brand Programmable Controllers, so this discussion focuses on integration with Rockwell Automation ControlLogix and Compact Logix PLCs

when the Forward Open is issued. A configuration wizard exists to walk the user through configuration of the adapter. AOP program adapters are significantly easier for end users to integrate than other adapters.

- **Adapters from Close Partners EDS Files** – This is the "Gold Standard" for adapters. These devices work identically to Rockwell devices. They are known devices, you can simply use the pulldown menu in the Logix programming tool to select one of these devices, just like the devices manufactured by Rockwell Automation. Note that only the closest of close partners to Rockwell get this service.

EDS File Structure

An EDS file follows a specific structure and is organized by sections where each section provides common information about the device. The sections in an EDS include (see Figure 35):

File Section – Administers the EDS file. Sometimes the URL keyword provides a link to a website where the latest version of the EDS can be found.

Device Section – Provides keying information that matches the EDS to a particular device revision. Four attributes of the Identity object (Vendor ID, Device Type, Product code and Product Revision) are used by network tools to match an EDS file to a device. Network tools generally do not connect to a device unless all four Identity object Parameters of the EDS file match the actual data in the device. Note that the Minor Revision number is not included in the match.

The ODVA recommends that the icon for the device be specified in this section so that users can have a graphical way of distinguishing devices.

Device Classification Section – Classifies the EDS for an EtherNet/IP network. The Device Classification Section is required for all EtherNet/IP devices.

Connection Manager Section – Identifies the CIP connections that are available in the device. This section indicates to the EtherNet/IP Scanner the triggers and transports available in the device. If a device supports multiple connections, then every connection must be detailed in this section.

Only connections that are specified in this section can be used in an EDS-based configuration tool.

Assembly, Params and ParamClass section – These sections are filled in as needed. For values that are limited to a defined set of values, Enumeration can be used to specify those values. Value ranges can be specified here also for Configurable parameters.

Capacity Section – This section indicates the number of connections available in the device and the connection speeds.

Port Section – This section describes the Ethernet port. It is only applicable to devices that perform CIP routing. It is unnecessary for devices containing a single CIP port.

Configuration Data

Various configuration capabilities are provided using EDS files. Some of them pertain to configuring individual attributes and parameters. Others pertain to the Configuration Data Block. The Configuration Data Block is a series of data words that are downloaded to a device on power up or reset. These blocks provide one-time initialization of an EtherNet/IP device.

Some of the ways EDS files can be used for configuration include:

Configuration Parameters – End users can configure a device using the enumeration, ranges and other limitations imposed on a configuration parameter by the EDS file.

Configuration Data Block Support – The Configuration Data Block can be loaded and later downloaded to the device in the Forward Open command.

Module Discovery – EDS files enable searching for their devices on the local network. When found, they can be easily loaded to a scanner's device list.

EDS File Tools

The ODVA no longer provides an EDS File Check tool to validate EDS file configuration. The ODVA now provides a tool called EZ-EDS™ for EDS creation and maintenance. It reduces the time to create EDSs through interactive menus. It provides support for all current EDS constructs with extension capabilities for any additional sections or keywords that might be added in the future. Using this utility, the EDS file construction process is no longer a lengthy, iterative process, but a straightforward task that can be completed quickly and effectively.

You can get the EZ-EDS tool directly from the ODVA on the ODVA website.

THE EVERYMAN'S GUIDE TO ETHERNET/IP

```
$ EZ-EDS Version 3.11 (Beta, May-09, 2012) Generated Electronic Data Sheet
[File]
    DescText = "ABC-FX100 EDS";
    CreateDate = 01-19-2012;
    CreateTime = 09:00:00;
    ModDate = 03-08-2016;
    ModTime = 07:39:28;
    Revision = 1.3;
    HomeURL = "http://www.COMPANY.com";

[Device]
    VendCode = 50;
    VendName = "Company name";
    ProdType = 12;
    ProdTypeStr = "Communications Adapter";
    ProdCode = 1456;          $ = 0x32CA
    MajRev = 1;
    MinRev = 15;
    ProdName = "EAW-PH12 EtherNet/IP 2-Port Remote I/O";
    Catalog = "EAW-PH12";
    Icon = "eaw_ph12.ico";

[Device Classification]
    Class1 = EtherNetIP;
            ;              $ decimal places not used

[Assembly]
    Revision = 2;
    MaxInst = 2;
    Number_Of_Static_Instances = 2;
    Max_Number_Of_Dynamic_Instances = 0;
    Class_Attributes =
        1,
        2,
        3;
    Instance_Attributes = 3;
    Class_Services = 0x0E;
    Instance_Services =
        0x0E,
        0x10;
    Assem100 =
        "Output",
        ,
        4,
        0x0000,
        "",
        8,Param100,
        8,Param100,
        8,Param100,
        8,Param100;

[Connection Manager]
    Revision = 1;
    MaxInst = 1;
    Number_Of_Static_Instances = 1;
    Max_Number_Of_Dynamic_Instances = 0;
    Connection1 =
        0x04030002,           $ trigger & transport
                    $   0-15  = supported transport classes (class 1)
                    $   16    = cyclic (1 = supported)
```

```
            $   17       = change of state (1 = supported)
            $   18       = on demand (1 = supported)
            $   19-23    = reserved (must be zero)
            $   24       = listen only (0 = not supported)
            $   25       = input only (1 = supported)
            $   26       = exclusive owner (1 = supported)
            $   27       = redundant (0 = not supported)
            $   28-30    = reserved (must be zero)
            $   31       = server 1
    0x44640405,       $ point/multicast & priority & realtime format
            $   0        = O=>T fixed (1 = supported)
            $   1        = O=>T variable (0 = not supported)
            $   2        = T=>O fixed (1 = supported)
            $   3        = T=>O variable (0 = not supported)
            $   4-7      = reserved (must be zero)
            $   8-11     = O=>T header
            $   12-15    = T=>O header
            $   16-19    = O=>T point-to-point
            $   20-23    = T=>O point-to-point or multicast
            $   24-27    = O=>T low, or high, or scheduled
            $   28-31    = T=>O low, or high, or scheduled
    Param4,Param2,Assem100, $ O=>T RPI,Size,Format
    Param3,Param1,Assem101, $ T=>O RPI,Size,Format
    ",                 $ config part 1 (dynamic assemblies)
    ",                 $ config part 2 (module configuration)
    "I/O Connection (SINT-Format)",  $ connection name
    "",             $ Help string
    "20 04 24 66 2C 64 2C 65";  $ path

[Port]
    Port1 =
        TCP,          $ port type name
        "TCP/IP",       $ name of port
        "20 F5 24 01",   $ instance one of the TCP/IP interface object
        2;           $ port number

[Capacity]
    MaxIOConnections = 2;
    MaxMsgConnections = 6;
    MaxConsumersPerMcast = 32;
    TSpec1 = TxRx, 2, 1000;
    TSpec2 =
        TxRx,
        64,
        $ to be verified
        500;
[TCP/IP Interface Class]
    Revision = 4;
    MaxInst = 1;
    Number_Of_Static_Instances = 1;
    Max_Number_Of_Dynamic_Instances = 0;
    Class_Attributes =
        1,
        2,
        3;
    Class_Services =
        0x0E,
        0x01;
    Instance_Services =
        0x0E,
```

THE EVERYMAN'S GUIDE TO ETHERNET/IP

```
            0x01,
            0x10,
            0x02;
         ENetQCT1 =
            350,
            50;              $ 350ms Ready for Connection time
                             $ 50ms Accumulated CIP Connection Time
    [Ethernet Link Class]
         Revision = 3;
         MaxInst = 3;
         Number_Of_Static_Instances = 3;
         Max_Number_Of_Dynamic_Instances = 0;
         Class_Attributes =
            1,
            2,
            3;
         Class_Services =
            0x0E,
            0x01;
         Instance_Services =
            0x0E,
            0x01,
            0x10;
         InterfaceLabel1 = "X1";
         InterfaceLabel2 = "X2";
         InterfaceLabel3 = "Port0-internal";
         InterfaceType1 = 2;
         InterfaceType2 = 2;
         InterfaceType3 = 1;
    $ End of File
```

Figure 35 - Sample EDS File (partial)

CIP EXTENSIONS

EtherNet/IP, ControlNet and DeviceNet are not the only technologies supported by CIP. There are CIP extensions for other technologies used on the factory floor. These extensions include Device Level Ring (DLR), CIP Safety™, CIP Motion™, CIP Sync™ and CIP Energy™.

DEVICE LEVEL RING (DLR)

Most manufacturers use the traditional star topology for their Ethernet networks. It's what is used in the office and is very familiar. Though used a lot, it's not the best topology for a machine network as the switch in the middle of the star becomes a single point of failure.

As the cost of switches decreased and it became feasible to embed a switch into a device, many users switched to a linear topology. A linear topology with devices daisy-chained to each other like the old RS485 networks was attractive – no switches to buy and a simpler physical layout. The devices became a little more expensive as they needed both a switch and two Ethernet jacks – one for input and one for output. The deficiency of the linear topology though was increased risk. A break in the line killed the entire network. Over the last ten years, there has been a growing trend to connect a number of EtherNet/IP devices in a linear fashion (line technology) with the last unit connected back to the first unit forming a ring of devices.

Device Level Ring (DLR) is a special layer 2 protocol that provides network redundancy, built-in fault detection and network fault resolution, all within a few milliseconds. It is very important to note that DLR is a layer 2 technology – the Ethernet Data Link layer. That means that DLR is not specifically a CIP or EtherNet/IP technology. DLR can coexist with most any Ethernet communications protocol

THE EVERYMAN'S GUIDE TO ETHERNET/IP

including PROFINET IO or Modbus TCP. DLR is typically built-into the lowest level Ethernet hardware access – the media or physical Ethernet interface with a software or hardware extension.

On a DLR-enabled network there are three kinds of devices: Ring Supervisors, Ring Nodes and non-DLR devices. Ring Supervisors are the ring network managers. Ring Supervisors monitor the network for device and network cable failures. Ring Supervisors send out messages to the network to determine status, issue reconfiguration messages to reconfigure the network on a failure and stop messages from circulating through the ring multiple times. Ring Nodes are simply nodes that have DLR technology and communicate with the Ring Supervisor. Non-DLR devices are nodes that do not have the DLR technology. These nodes can function normally on a DLR network but can't participate in the actions to recover from network faults. It is common practice to never group non-DLR nodes together as a Ring Supervisor would be unable to detect exactly where a network fault occurred within a group of non-DLR nodes. If you need to add a non-DLR node to a DLR network, consider a DLR 3-port Ethernet switch external device to turn your non-DLR node into a DLR node.

Another category of device is the backup Ring Supervisor. Backup Ring Supervisors standby in case of failure of the Ring Supervisor. There can be multiple backup Ring Supervisors with the MACID of the backup Ring Supervisors determining the order of precedence when the current Ring Supervisor fails.

There are two types of frames used by the DLR protocol: Beacon frames and Announce frames. Beacon frames are sent in both directions from the Ring Supervisor every 400 μs. Ring Supervisors detect faults when a Beacon does not return to the Ring Supervisor's other port. There are two classes of DLR devices: Beacon-based and Announce-based, which transmit Beacon frames and Announce frames, respectively. Both frames send essentially the same information. Devices that don't have the capability to process Beacon frames every 400 μs support Announce frames, which are transmitted must less frequently.

EtherNet/IP provides a DLR object (Object Class 47_{HEX}) to support the DLR protocol. The DLR object includes a number of attributes including these:

- A Network Technology attribute to identify if the network is in linear or ring mode
- A Network Status attribute to identify if the network is in normal, faulted or other non-normal operation mode

- An Active Supervisor attribute to identify which node is currently the Ring Supervisor. Application layer protocols use the DLR object to monitor the status of the ring.

CIP SAFETY™

More critical than consistent, reliable performance of a manufacturing process is safe operation. Manufacturers devote an extraordinary amount of time, energy and investment to developing processes and procedures and procuring equipment that will offer safe operation.

CIP provides an extension to EtherNet/IP that enables manufacturers to build safety into their process and achieve Safety Integrity Level (SIL) 3 operation. Safety Integrity Levels are a certified mechanism documented in IEC standard 61508 that specify how to achieve fail-safe operation. Safety systems are rated by SIL level, with higher levels having a smaller probability of failure. SIL 3 systems have a very low probability of failure.

CIP Safety is a mechanism for providing SIL 3 fail-safe operation using nodes designed for safe operation like I/O blocks, interlock switches, light curtains and controllers and as certified by TÜV Rheinland. CIP Safety devices have been working in the field since 2005, and according to the 2011 World Market for Industry Networking study by IMS Research, CIP Safety is one of the most implemented safety networking systems and has also been adopted by other trade associations including SERCOS International as the only safety protocol for use on SERCOS III networks.

Users may mix safety devices with standard devices on the same network using CIP Safety. Nonredundant hardware can be used for communication as safety application layer extensions do not rely on the integrity of the underlying standard CIP services and data link layers.

CIP SYNC™ AND CIP MOTION™

Motion is integral to every manufacturing system. Every process moves something. And in a large number of these processes, that movement, that motion must be coordinated for optimal results. In papermaking, for example, massive webs of paper are moved from the drum to a roller. Any discrepancy between the drive motors results in a tear and a massive loss of production. Coordinated motion can be as simple as two motors running concurrently to multi-axis, multi-node control of complex systems.

Analog signals were originally used to coordinate motion, but analog control gets difficult as the number of nodes, distances and complexity of the application increases. Using Ethernet, systems were created over the years that used the Ethernet infrastructure to coordinate motion. These systems relied on special software in switches, routers and end devices to enable time slicing. In a time-slicing system, every node gets a slice of time to get its reference information and deliver its feedback. These systems are more complex and costlier as every device requires special hardware and software.

A more standardized way to implement motion systems in an EtherNet/IP system is to use CIP Sync and CIP Motion. CIP Sync uses the IEEE 1588 standard, known as the Precision Time Protocol (PTP), to very precisely synchronize clocks in devices across the network. CIP Sync automatically corrects for infrastructure delays and network latencies. With CIP Sync and PTP, clocks in distributed devices are synchronized to an accuracy of hundreds of nanoseconds.

Once all the motion devices on a network are using the same time clocks, CIP Motion is able to deliver commands and data with timestamps. The timestamps enable devices to execute motion changes on the schedule received from the master motion controller. Controllers, instead of focusing on the time slices, can simply plan a motion based on a series of time-sequenced operations. This enables much simpler operation.

Multi-axis motion typically synchronizes on an event-basis, which requires scheduled, absolute hard delivery of time-critical data across the network with less than 1 μs of Jitter. CIP Motion is one of the few solutions that delivers deterministic, real-time, closed loop motion control for multi-axis systems over standard, unmodified Ethernet.

CIP ENERGY™

Energy usage is a critical resource in the United States and around the globe. Massive energy consumers that you might not expect include some of the most well-known fast food companies. Your average fast food enterprise running seven fryers at full power consumes the same amount of energy as a steel plant on a per square foot basis. Manufacturing systems, as you might appreciate, aren't known for efficient energy use. About one third of all energy use in the US is from manufacturing plants,[22] with motors consuming a significant majority of that energy. Monitoring and controlling all this energy not only makes financial sense but has nationwide impact as it

---

[22] Energy Information Agency statistics

drives our future needs for coal, gas and nuclear resources.

Unfortunately, energy usage is one of the most difficult costs for a manufacturer to manage. Many legacy devices have no mechanisms to report energy data, and a lot of them don't even measure it. Newer devices use various mechanisms and report the data in different ways. A comprehensive energy management system that can resolve these inconsistencies has been deemed too costly by many manufacturers. CIP Energy is an attempt to coalesce this important data into a standardized data set where it can be accessed, monitored and used to conserve energy resources.

CIP Energy is an extension to the standard CIP Object Model structure that provides a family of objects, attributes and services that standardizes how energy data is gathered and reported over CIP networks. With CIP Energy, operations managers can receive energy data over a CIP network in real time, creating baseline energy profiles. With that baseline, production energy usage can be optimized, and problems can be diagnosed for specific processes. The largest energy consuming equipment in a plant can be replaced with new equipment having a smaller energy footprint. By matching energy usage with plant operations cycles, opportunities can be found to reduce energy usage of idle equipment.

Using CIP Energy across an entire manufacturing enterprise enables manufacturers to quickly and easily aggregate energy data from many different devices. With that data, actionable information can be used to manage energy consumption, reduce operating expenses and preserve precious resources.

THE ODVA

Trade organizations are always helpful, both to the novice and the experienced professional. The annual events they put on are usually informative. Networking with other professionals using the same technology has its benefits. And if you're a novice, it is the best place to find seasoned professionals who can give you the one or two pieces of advice that will save you countless hours of frustration.

The Open Device Vendor Association, now known as simply the ODVA, is a global association of automation companies creating products for the Common Industrial Protocol (CIP). Unlike some other manufacturing automation trade associations, ODVA membership is restricted to companies developing products using CIP technology. To qualify for membership, a company must manufacture hardware and/or software products that either integrate CIP or are designed to enhance the implementation, operation or support of products that integrate CIP.

The ODVA is a non-profit entity, founded in 1995 as the Open DeviceNet Vendor Association by four of the world's leading automation companies: Rockwell Automation, Omron, Cutler-Hammer and Square D. Its original mission was to promote and support DeviceNet technology. In 2001, the Open DeviceNet Vendor Association signed a joint technology cooperation agreement with ControlNet International Ltd to coordinate and promote the ControlNet standard.

In 2009, the Open DeviceNet Vendor Association assumed all the rights and responsibilities for EtherNet/IP and changed its legal name to ODVA, Inc. Today, ODVA manages a technology portfolio that includes DeviceNet, EtherNet/IP, ControlNet, the Common Industrial Protocol (CIP), and other related technologies like CIP Sync and CIP Safety.

The ODVA focuses on the following areas:

- **Technology**: The ODVA, through its Technical Review Board (TRB) and numerous Special Interest Groups (SIGs), is the arbiter of what CIP is, how it works and the standards for products. The organization works to not only promote the technology, but to make sure it is implemented correctly and in a fashion that promotes interoperability among supplier products. The ODVA is a clearing house for products, technical information, tools and services that are useful to CIP members and the automation community. Visitors to the organization's website can find a list of resources to assist them in the development and use of the ODVA technology portfolio.
- **Promotion:** The ODVA promotes the adoption of CIP technology by publishing a newsletter highlighting vendor products and projects, exhibiting at trade shows where technology users gather, and holding seminars and other events to inform and train people on its technologies.
- **Standards:** The ODVA establishes compliance criteria and conformance and compatibility standards and performs product certification at its lab in Ann Arbor, MI, to ensure that CIP products conform to approved technology standards. Products attaining approval are granted a Declaration of Conformity (DOC).

There are significant advantages for automation product vendors to joining the ODVA and developing products using CIP technologies:

1. Product listings on the ODVA website. Vendors can list the features and benefits of products. End users, distributors and integrators can quickly and easily find them.
2. Communications that provide interesting information on member products, seminars and latest developments in the industry.
3. Cost effective technical training programs.
4. The opportunity to participate in technical committees that influence the future of the technology.

Members of the ODVA can use the member logo (Figure 36) on their company literature to signify their participation in the global association of automation companies creating products for the Common Industrial Protocol.

THE EVERYMAN'S GUIDE TO ETHERNET/IP

Figure 36 – ODVA Membership Logo

ODVA provides a legal framework for its members, who are in fact competitors, to collaborate in a way that complies with antitrust and competition laws as they pertain to trade and standards development organizations. ODVA ensures that the intellectual property created through member collaboration by developing work products, such as its technical specifications, is vendor-neutral and open.

ODVA is headquartered in Ann Arbor, Michigan, USA, and incorporated in the State of Wisconsin, USA.

ODVA, Inc.
4220 Varsity Dr,
Ann Arbor, MI 48108
+1 734 975 8840
+1 734 922 0027 (F)
odva@odva.org
www.odva.org

ETHERNET/IP AND THE IOT

There have been more seminars, webinars and words written about the Internet of Things than any other topic of the last 40 years. What some people forget is that IoT is not new to us in the automation world. We've always been pulling data out of factory floor systems and moving it into business systems and databases in the enterprise.

It started out pretty crudely with operators writing operating data on sheets that were later keypunched and loaded onto magnetic tape. Once a week, the tapes were loaded, reports were run on that data and dropped on people's desks around the factory for review. Some were, of course, "mailed" to corporate for their analysis.

Eventually, many factories had people walk around the factory floor collecting memory sticks out of devices, carrying them back to their desks, uploading the data, converting it and then, finally, loading it into a database. It was pretty rudimentary, but it was sort of effective because only the most important data was collected.

It wasn't always that crude. A lot of the larger manufacturing companies spent massive sums to build proprietary systems that could move production, quality and other operational data from their manufacturing systems to an MES or ERP system automatically. Those systems were complicated, slow and a nightmare to maintain, but they worked.

IoT in manufacturing is just the latest evolution of what we've always been trying to do. What's new is that we now have technologies that make it easier than ever to accomplish. But that, in fact, is our biggest problem. It is now so very easy to move data to the Cloud that we have all sorts of manufacturers putting data there with no idea why they're doing it. In many cases, there is no business model, no plan to monetize the data, no thought about ROI – no plan at all for how to use the data. Many engineers have these hammers in their toolbox (MQTT, AMQP, OPC UA) so they just start swinging away. There are statistics from Cisco and others that indicate that as little as one in four IoT projects achieve any ROI.

In fairness, C-suite people are pushing these engineering staff to implement IoT as they are bombarded by consultants promising huge payoffs and hordes of vendors offering painless, all-inclusive solutions. These vendors portray that they can pull data from anything, as fast as you could possibly want, in any industry with no deployment cost or effort on your part.

While many of these vendors oversell their IoT solutions, there is some truth in the assertion that in a highly competitive, global environment, factories must become faster, more efficient and automated. The applications offered in the Cloud by Amazon (AWS) and Microsoft (Azure) offer manufacturers tools to increase productivity and efficiency. In fact, there are an amazing array of tools on offer: compute tools, storage tools, database tools, presentation tools and network management tools. The list is almost endless and seems to grow by the day.

And to take advantage of those tools, data – all of it, even the data stored in EtherNet/IP devices – must be made available to these Cloud-based applications. But, if it's valuable to transfer CIP/IP attributes between scanner and adapters and Cloud-based applications, how do we accomplish it?

Before we discuss that, there is one fact that must be made clear:

Modbus TCP and PROFINET IO are not IoT transport protocols!

It's unclear where the idea that Ethernet I/O protocols could be used as IoT protocols started, but it probably started with some trade association. The facts are clear. EtherNet/IP, PROFINET IO and to a lesser extent, Modbus TCP are really perfect as they are. They move I/O data from end devices into controllers quite well. Well enough that I'd suspect that the majority of the world's manufacturing systems use these protocols.

These trade associations and the vendors in them have done a great job of defining the right combination of transports, encodings, messaging, and data connections that solve the I/O communications problem in manufacturing facilities all over the world.

But there are now some misguided proponents of these technologies that want to use EtherNet/IP, PROFINET IO and Modbus TCP to move IoT data. That's just an indefensible position. Those technologies move I/O data that's very strictly defined between an end device and a controller. Outputs move from the controller to the end device. Inputs move from the end device to the controller. And they move continuously, over and over, usually every 1 ms or 10 ms.

That's not what you want in the IoT world for a number of reasons. First, the data can't be that strictly defined. There are lots of different kinds of data to move in the IoT world and it changes from moment to moment. Second, there's no security associated with these technologies. They're working on that, but when it will get to market is an unknown. Third, the data is intermittent. I don't want my garage door system eating up bandwidth telling me it's closed every 10 msec. Fourth, these technologies are not designed to work over the internet. They could, but they'd eat up all the bandwidth. Fifth, the data representation doesn't work for IoT devices. In PROFINET IO, every end device must look like an I/O rack. It's bad enough that I am required to make a barcode scanner look like an I/O rack to a PROFINET IO network Controller, but would I really want to make my water softener look like an I/O rack too?

Use the Ethernet application layer protocols (EtherNet/IP, Modbus TCP and PROFINET IO) in applications where it's best suited, and that's not in IoT applications. There are better technologies for moving data in and out of the Cloud. Some of them work well with programmable controllers and other devices that may also support EtherNet/IP. EtherNet/IP and the other Ethernet application layer protocols work well in parallel with IoT technologies, not as a replacement for them.

MOVING ETHERNET/IP DATA TO THE CLOUD

But if it's valuable to use EtherNet/IP and other Ethernet protocols in parallel with IoT technologies, what is the best way to connect an EtherNet/IP device to one of the numerous Cloud providers or your own provide Cloud? There are lots of ways to go about this and many technology providers that have solutions for you.

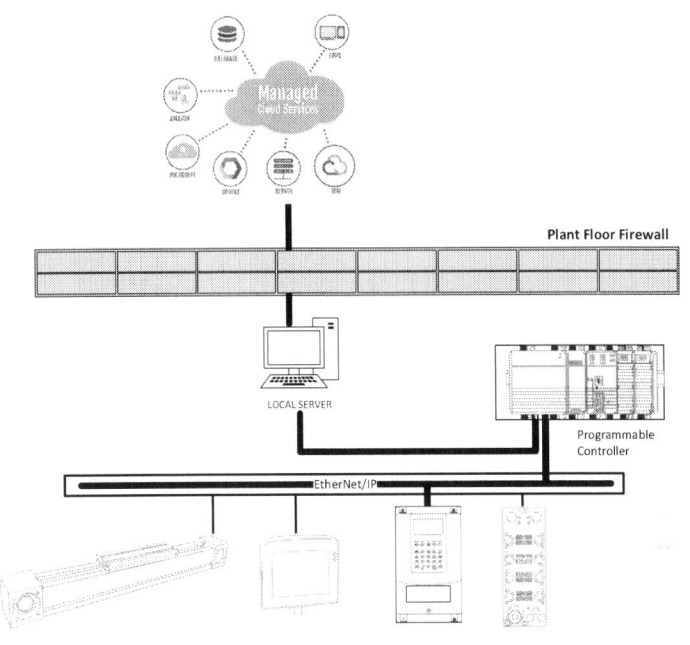

Figure 37 – Traditional Architecture to Move Data to the Cloud

One way, the traditional way, is presented in Figure 37. All or nearly all programmable controllers have an OPC driver that can move data into a local PC server. Once in the PC server, there are many ways to take OPC data and deliver it to a Cloud service. The major disadvantage of this solution is that it relies on an additional server to provide the Cloud interface, and the data must be mapped from the programmable controller data space to OPC variables and then to a Cloud database. That sort of mapping has always been problematic and unreliable, but many manufacturers have relied on it for a very long time.

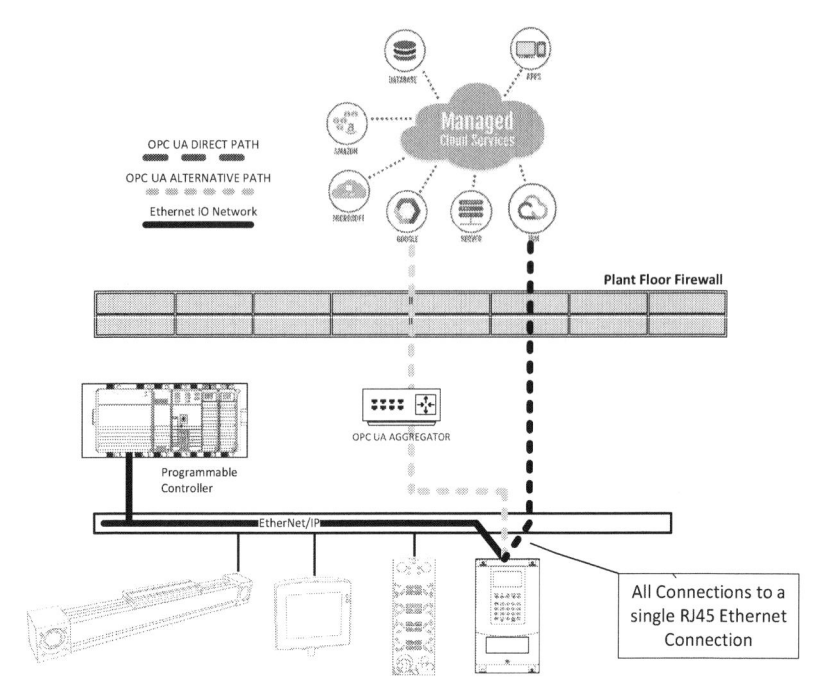

Figure 38 – EtherNet/IP and OPC UA Cloud Communication

Another mechanism (Figure 38) is to enable specific devices on the factory floor with both an I/O protocol like EtherNet/IP and a Cloud friendly technology like OPC UA. In this architecture, these devices use traditional EtherNet/IP communications for control while also providing data to the Enterprise and/or the Cloud over OPC UA. If you can find a device with dual capabilities, you can have the best of both worlds: traditional, programmable controller control while also capturing data for historians, application programs, HMI servers and other applications in the Cloud. Negotiating firewall issues and making appropriate connections can be problematic in this solution depending on the OPC UA capabilities of your factory floor devices.

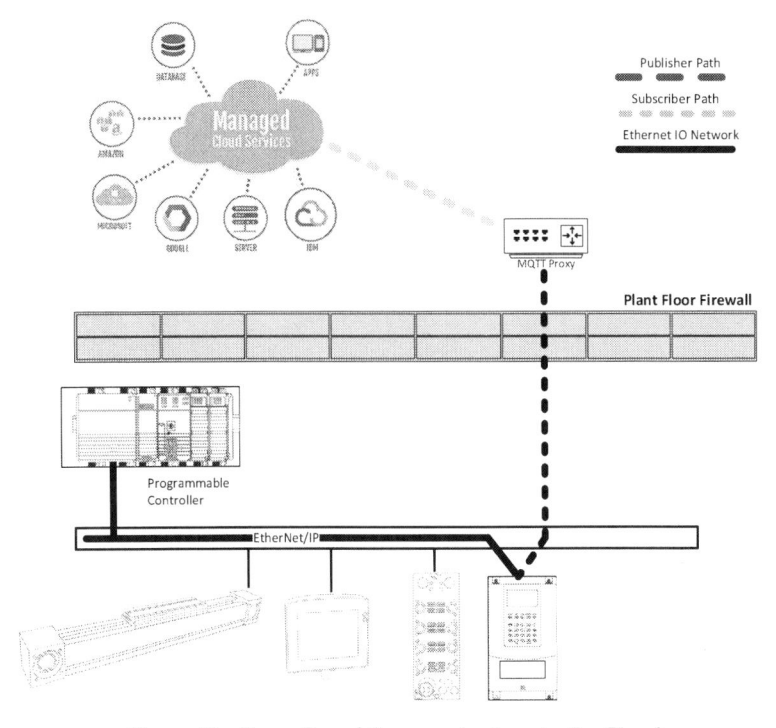

Figure 39 – Proxy Based Communications to the Cloud

Other manufacturers are using simple MQTT (Message Queuing Telemetry Transport) communication to move data to the Cloud (Figure 39). This has similar architecture to the previous solution but with the simplicity of the lightweight MQTT protocol. Again, firewall issues and contextual issues regarding the identification of the MQTT data will need to be solved to implement an architecture like this.

It is inevitable that the data in your EtherNet/IP scanners and adapters will be needed in the Cloud. It's just a matter of time, and there are lots of ways to solve that problem. Three of them are presented here. You might have other solutions. The point to remember though is that EtherNet/IP (and PROFINET IO and Modbus TCP) are all just I/O protocols and that you'll need other technology to get the advantages of the analytic, database, presentation and other tools that are available in Cloud systems. EtherNet/IP is not that technology.

JOHN S. RINALDI

ETHERNET/IP VS. PROFINET

If you have some gray on your head, you probably remember one or more technology fights in your career. One of the more famous ones that you may remember was the commercial war between VHS and Betamax. Sony released its Betamax format for consumer recording (Analog mind you, <u>not</u> Digital) of video shortly after the first system was created by Phillips in the early 1970s. A competing standard quickly entered the picture, the VHS format from JVC, followed by other, minor players.

Sony had hoped that its 1974 prototype videotape recording system ("Beta") would be adopted by other electronics manufacturers. They believed that their superior recording resolution and adoption of a single standard would be best for the consumer. That was a good thought, but, of course, other manufacturers didn't agree. By 1981, VHS captured the market due to its superior recording time and lower cost, despite the better quality of the Betamax systems.

This kind of technical fight where the higher quality system loses isn't atypical. The Betamax/VHS war is illustrative of what has happened in fieldbus technology over the last 25 years. One vendor introduces a technology and another major vendor, afraid of the implications from adopting a competitor's technology, feels pressure to introduce a competing technology. The vendor partners, trade associations and distributors all go to war. Brochures, trade journal articles, training sessions and trade events are used as artillery to disparage the other technology.

This has happened a number of times over the years with all sorts of technologies. It's been most prominent in the Profibus DP vs. DeviceNet battle, the EtherNet/IP vs. PROFINET IO war and with Modbus TCP vs. everything else. And no one really wins. And as we see with VHS, now that we all use streaming services, we eventually wonder why we fought the battle at all.

The question often asked about manufacturing networks is, "What's the best fieldbus technology?" The question is entirely irrelevant because there is no "best" technology. Though you can describe the differences between one networking technology and another, no one ever purchased a manufacturing system, a programmable controller or an I/O system because of the capabilities of the fieldbus system. They selected the programmable controller and the related components that met the price, performance and reliability of the application and lived with whatever fieldbus technology was supported on that controller.

In Allen-Bradley shops, they used DeviceNet, EtherNet/IP (and Modbus when they had no other choice). In Siemens shops they used Profibus DP and PROFINET IO (and Modbus when they had no other choice). Mixing systems was avoided as much as possible as the costs of spares, training and support for multiple systems was often prohibitive.

The other question that is often asked is "What's the difference between the technologies?" Not a bad question as an intellectual exercise. Table 12 illustrates some of the major differences.

	EtherNet/IP	PROFINET
Largest Supplier	Allen-Bradley	Siemens
Trade Association	ODVA	PROFINET International
Standard Ethernet	YES	YES
TCP/IP Integration	Standard TCP and UDP	Starndard UDP Requires special integration to intercept Layer 2 messages
Data Model	Object Model with objects, instances and attributes	Racks, Slots, Modules, Channels and Points
Cyclic Messaging	I/O Messaging	I/O Messaging
Acyclic Messaging	Explicit Messaging	Read/Write Indices
Motion Control Support	CIP Sync and CIP Motion	PROFINET Motion
Topology	Standard: Star and tree topology Line and ring topologies are supported	Standard: Star and tree topology Line and ring topologies are supported
Sever Description File	EDS (Electronic Data Sheet)	GSDML (General Station Description Markup Language)
Diagnostic Support	Vendor Supplied	Required
Wireless Support	802.11	802.11
Certification	Required	Required
Industry Support	Mostly Discrete	Discrete plus some process and more motion than EtherNet/IP
Safety Support	CIP Safety	PROFIsafe
Typical I/O Performance	10 msec updates	1 msec updates
Device Replacement	Yes, as long as the IP is configured the same	Yes, natively supported when topology diagnostics are enabled
Standard Unmodified Ethernet	Yes	Yes (except for isochronous real time)

Table 12 – EtherNet/IP vs. PROFINET Comparison

What you'll notice in Table 12 is that the two technologies are very similar. There are a few differences around the edges in the key characteristics listed in this table. What's really not clear from this list is that PROFINET IO is a higher performance technology with a vast set of additional functionalities. It includes advanced diagnostic capabilities, features like SNMP (Simple Network Management Protocol) and more that isn't typically required in a lot of applications. That additional functionality and higher performance comes with increased cost and time to develop it, test it, integrate it, maintain it and certify it. For example, PROFINET IO generally requires ten times more resources (code space, RAM, engineering labor) to get a device certified than EtherNet/IP does.

There is no winner in the contest between EtherNet/IP and PROFINET IO but there are casualties. And just like the Betamax/VHS war, at some point, another technology may, in the not too distant future, make this battle irrelevant.

JOHN S. RINALDI

ETHERNET/IP AND OPC UA

It's hard to imagine anyone working in the automation industry who doesn't know about OPC. It's been unarguably one of the most successful technologies to ever hit the factory floor. Over the last twenty years, OPC has provided a standardized way to move data from some factory floor device into your Windows application.

OPC was based on Microsoft COM (Distributed COM), a basic Microsoft Windows transport technology that provides the standard communications architecture between tasks running in Windows. OPC uses it to move data from the OPC Server, the interface to your factory floor device and an OPC Client like a Windows HMI, Historian or other Windows application. Tens of thousands of these clients and servers have been deployed for thousands of devices over the years, and OPC has found its way into most factory floor architectures.

Now OPC is being renamed OPC Classic and being replace by OPC UA (Unified Architecture). It vastly expands the scope of what OPC can do and vastly increases its capabilities. If you're not familiar with this technology, here's a quick summary of what you need to know:

1. OPC UA IS NOT A PROTOCOL

It's a common misconception that OPC UA is just another protocol. That couldn't be farther from the truth.

A computer protocol is a set of rules that govern the transfer of data from one computer to another. The protocol specification governs how the data is to appear on the wire, how the conversation between the two computers starts and ends, and what the message structure is between the two computers. It is very rigorous and unforgiving. Now, even though OPC UA also specifies the rules for communication between computers, its vision is more than just moving some arbitrary data from one computer to another. OPC UA is about complete interoperability.

It is an architecture that systematizes how to model data, model systems, model machines and model entire plants. You can model anything in OPC UA.

It is a technology that allows users to customize how data is organized and how information about that data is reported. Notifications on user-selected events can be made on criteria chosen by the user, including by-exception.

It is a scalable technology that can be deployed on small embedded devices and larger servers. It functions as well in an IT database application as on a recipe management system, on a factory floor, or a maintenance application in a Building Automation system.

It is a technology that provides layers of security that include authorization, authentication and auditing. The security level can be chosen at runtime and tailored to the needs of the application or plant environment.

OPC UA is a systems architecture that promotes interoperability between all types and manner of systems in various kinds of applications.

2. OPC UA IS THE SUCCESSOR TO OPC (NOW OPC CLASSIC)

OPC UA solves the deficiencies and limitations of OPC Classic. In today's world, we need to move data between all sorts of embedded devices, some with specialized operating systems and software, and enterprise/Internet systems. OPC Classic was never designed for that.

OPC Classic is dependent upon Microsoft – It's built around DCOM (Distributed COM), a technology that is, if not obsolete, certainly being de-emphasized by Microsoft.

OPC Classic has no support for sophisticated data models – It lacks the ability to adequately represent the kinds of relationships, information, objects and interactions among systems that are important in today's connected world.

OPC Classic is vulnerable – Microsoft platforms that support COM and DCOM are vulnerable to sophisticated attacks from all sorts of viruses and malware.

UA is the first communication technology built specifically to live in that "no man's land" where data must traverse firewalls, specialized platforms and security barriers to arrive at a place where that data can be turned into information.

3. OPC UA SUPPORTS THE CLIENT–SERVER ARCHITECTURE

We all very familiar with technologies that have some superior/subordinate relationship. Modbus is one that everyone knows. There is a Modbus RTU Master and a Modbus RTU Slave. A Modbus TCP Client and a Modbus TCP Server. It's the same for EtherNet/IP, only with EtherNet/IP, the terms are EtherNet/IP Scanner and EtherNet/IP Adapter. The same for BACnet: there is a BACnet Client and a BACnet Server.

The difference between these technologies and OPC UA is that in all the familiar Industrial and Building Automation protocols, the client or master somehow takes ownership of the server or slave device. In most of these technologies, once a client takes ownership of a slave, no other client or master can access it.

That's not true of OPC UA Clients and Servers. In OPC UA, a server can be configured to accept connections with one, two or any number of clients. A client device can connect and access the data in any number of servers. It's much more of a peer relationship in OPC UA, though, like other technologies, servers simply respond to requests from clients and never initiate communications.

Another unusual and interesting aspect of this relationship is that in OPC UA, a server device can allow a client to dynamically discover what level of interoperability it supports, what services it offers, what transports are available, what security levels are supported, and even the type definitions for data types and object types. A server in OPC UA is a much more sophisticated device than in many of the technologies you've worked with in the past.

4. OPC UA IS A PLATFORM-INDEPENDENT AND EXTREMELY SCALABLE TECHNOLOGY

Unlike OPC Classic, OPC UA is designed from the ground up to be platform independent. Ethernet and some mechanism to know the current date/time are the only requirements for OPC UA. OPC UA is being deployed to everything from small chips with less than 64K of code space to large workstations with gigabytes of RAM.

All the components of OPC UA are designed to be scalable, including security, transports, the Information Model and its communication model. Several security models are available that support the level of security appropriate for the device's resources and processor bandwidth.

An encoding mechanism can be selected to provide ease of communication with IT systems (heavy RAM footprint and processor intensive), or one that provides fast, smaller message packets (light RAM footprint and less processor bandwidth).

The same scalability exists with the address space. An address space can be comprised of a few objects with a single variable or a sophisticated, complex set of interrelated objects.

Transports are often limited in most technologies, but not OPC UA. In OPC UA, a device may only support a single transport, or it may support a set of transports that allow for communications over various kinds of networks.

5. OPC UA INTEGRATES WELL WITH IT SYSTEMS

There is an ever-increasing emphasis on connecting the factory floor to the enterprise. Management, customers, and suppliers all want increased efficiency, productivity, higher quality, and the like. Government regulation is forcing much more data archiving. All these factors are increasing the need to connect factory floor systems to non-factory floor processes, enterprise applications, and Cloud services.

OPC UA is designed to be well-suited for this. Servers can support the transports used in many traditional IT type applications. Servers can connect with these IT applications using SOAP (Simple Object Access Protocol) or HTTP (Hypertext Transfer Protocol). HTTP is, in fact, the foundation of the data communication used by the World Wide Web.

OPC UA Servers can also support XML encoding, the encoding scheme used by many IT-type applications. It's likely that most servers in the factory floor automation space will not support XML encoding due to the large resources required to decode and encode XML. Instead, many servers in that space will support OPC UA Secure Conversation, a more efficient, binary encoding that uses more limited resources.

A key differentiator for OPC UA is that the mapping to the Transport Layer is totally independent of the OPC UA services, messaging, Information and Object Models. That way, if additional transports are defined in the future, the same OPC UA Information Model, Object Model and messaging services can be applied to that new transport. OPC UA truly is future-proof.

6. A SOPHISTICATED ADDRESS SPACE MODEL

The address space model for OPC UA is more sophisticated than EtherNet/IP, PROFINET IO, Modbus or any of the Industrial or Building Automation protocols. The fundamental component of an OPC UA address space is an element called a Node. A Node is described by its attributes (a set of characteristics) and interconnected to other nodes by its references or relationships with other nodes.

All nodes share common attributes with all other nodes in its Node Class. Every Node is an instantiation of its Node Class. Node Classes include the Variable Node Class for defining variables, the Reference Node Class for defining references to other nodes and the Object Type Node Class, which provides type information for Objects and sets of Objects.

The attributes of a Node include its Node Class, its Browse Name, its Display Name, its value and its *Node ID*. There are 22 possible attributes for a Node. A subset of those nodes is mandatory for each Node Class. For example, only Nodes of the Variable Node Class may instantiate the Value attribute, which contains the current data value.

7. OPC UA PROVIDES A TRUE INFORMATION MODEL

The ability of an object Node to have references to other object Nodes that further reference to other object Nodes to an unlimited degree provides the capability to form hierarchical relationships that can represent systems, processes and information - an Information Model.

An Information Model is nothing more than a logical representation of a physical process. An Information Model can represent something as tiny as a screw, a component of a process like a pump, or something as complex and large as an entire filling machine. The Information Model is simply a well-defined structure of information devoid of any details on how to access process variables, metadata, or anything else contained within it.

OPC UA provides a systemized way to create and document Information Models. An XML Schema serves as the mechanism for documenting and communicating Information Models. The OPC UA Foundation, trade associations and vendors are all able to develop and propose Information Models for specific devices and systems by describing them using the XML Schema defined by the OPC UA Foundation.

Though OPC UA is not the first organization to systemize the creation and documentation of Information Models, it is the first technology to provide the mechanism to load, transport and reference those Information Models in a running system. A client device that detects the use of an Information Model in a server device can access that model (through its URI) and use that model to access the Information Model in a server.

Many trade groups, including Gas and Oil, the BACnet Association, the PLCopen organization, and others, are using the Information Model processing of UA to define Information Models for their application domains and simply use UA for the standard transports, security and access to their data models.

8. OPC UA IS NOT A FACTORY FLOOR PROTOCOL LIKE ETHERNET/IP

I always cringe when I hear OPC UA compared to EtherNet/IP, PROFINET IO, or Modbus TCP. That wasn't the case when PROFINET IO came out. I could tell people that it was the equivalent of EtherNet/IP for Siemens controllers. Same kind of technology. Basically, the same kind of functionality. Easy to explain.

I can't do that for OPC UA. I could say that it's Web Services for automation systems. Or that it's SOA for automation systems, an even more arcane term. SOA is "Service Oriented Architecture," basically the same thing as Web Services. That's fine if you're an IT guy (or gal) and you understand those terms. You have some context.

But if you're a plant floor person, it's likely that even though you use Web Services (it's the plumbing for the internet), you don't know what that term means.

And it's just as likely that if you're a plant floor person, you also say, "Why do we need another protocol? Modbus TCP, EtherNet/IP and PROFINET IO work just fine." The answer is that it's not like EtherNet/IP, PROFINET IO or Modbus TCP. It's a completely new paradigm for plant floor communications. It's like trying to explain EtherNet/IP to a PLC programmer in 1982. With nothing to compare it to, it's impossible to understand.

In the automation world, the PLC networking paradigm is second nature. You have a PLC; it is a master kind of device, and it moves data in and out of slave devices. It uses very simple, transaction-type messaging or some kind of connected messaging. In either case, there is this buffer of output data in a thing called a Programmable Controller. There is a buffer of input data in a bunch of devices called servers, slaves or nodes. The buffer of input data moves to the Programmable Controller. The output data buffers move from the Programmable Controller to the devices. Rinse. Repeat. Forever. Done.

OPC UA really lives outside that paradigm. Well, actually, that's not true. OPC UA lives in parallel with that paradigm. It doesn't replace it. It extends it. Adds on to it. Brings it new functionality and creates new use cases and drives new applications. In the end, it increases productivity, enhances quality and lowers costs by providing not only more data, but also information – and the right kind of information – to the production, maintenance, and IT systems that need that information when they need it.

And that's the stuff that EtherNet/IP PROFINET IO, Modbus TCP and all the rest just can't do. OPC UA is a perfect complement to EtherNet/IP and these other technologies, it's not a replacement for them.

ETHERNET/IP AND DEVICENET

DeviceNet was a revolutionary technology when it was introduced in 1994. You should realize that prior to that, the only automation network that anyone could readily name was Modbus.[23] Even for devices that used proprietary protocols like chillers, everything was a request followed by a response over RS485 serial communications. Send the request, get the response, and move on to the next device request and response... ad infinitum. At the end of the device list, you started over. That's how everything on RS485 (and RS422) worked. As far as anyone knew, that was the only way to do manufacturing networking.

When DeviceNet was introduced, it was just like that scene from that old movie "The Day the Earth Stood Still." The Alien spaceship lands and no one knows what it is, what it means, and where we go from here. That's what the introduction of DeviceNet was like.

Most people were completely stunned by the idea of bit-wise arbitration. DeviceNet is built on CAN (Controller Area Networking), and bit-wise arbitration is the process in CAN where all devices with messages to send synchronously transmit their bits. That's unlike most technologies where one node talks, and all the others listen and wait for their turn to talk. In CAN, every device with a message to send synchronously transmits their bits and continues transmitting until they realize that a higher priority message is on the bus.

The key to that CAN (and DeviceNet) prioritization scheme is that after transmitting a bit, each node listens as that bit is reflected on the

---

[23] ControlNet was introduced around the same time, but few users adopted it. DeviceNet made a much bigger impact.

network. Since zero bits dominate one bits, anytime a node sends out a one and hears a zero reflected, it knows that there is some other node with more leading zeros in the message. Having more leading zero is how CAN messages are prioritized.

A node that doesn't hear its one reflected stops transmitting, so any time two or more devices are sending the same sequence of bits, the devices just continue sending bits. The highest priority device – the one sending a message with the most leading zeros – eventually becomes the last device to keep sending. The genius in this is that a node never needs to know anything about prioritization. Messages just transmit, completely oblivious to any bus conflict.

Bitwise arbitration enables DeviceNet to achieve 100% bus utilization since there is never any destructive bus contention. In 1994, that kind of arbitration along with 500K baud data rates and power on the bus made DeviceNet a truly revolutionary new technology unlike anything ever seen before.

But 1994 was a long time ago. Now, the direction for automation is clearly Ethernet (EtherNet/IP, PROFINET IO, Modbus TCP, OPC UA).

So where does DeviceNet fit in an Ethernet World?

We all know that there are hundreds of thousands of DeviceNet nodes all over the world. And we also know that DeviceNet is not going to be easily replaced in those applications. The technology works and it's well-understood and maintainable. It's also price competitive to Ethernet – especially when you figure in the cost of Ethernet switches.

DeviceNet is, in fact, perfect for applications where the topography is such that nodes are arrayed linearly. That's the DeviceNet sweet spot: nodes arrayed linearly like a conveyor line. Unless something changes, you can expect DeviceNet to be deployed in these kinds of applications well into the future.

DeviceNet also has the advantage that it is uses CIP exactly as EtherNet/IP does only over a CAN Physical Layer. Instead of the TCP/IP stack and Ethernet Physical Layer, DeviceNet uses CAN (Figure 40).

THE EVERYMAN'S GUIDE TO ETHERNET/IP

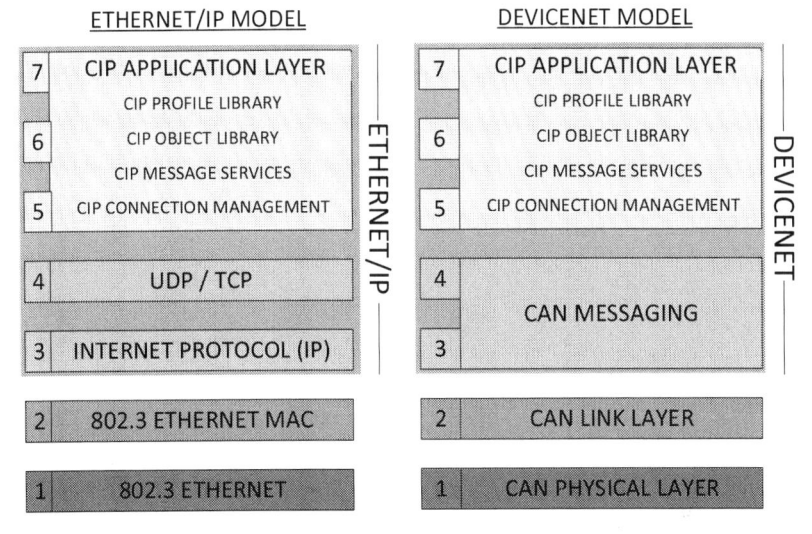

ETHERNET/IP MODEL		DEVICENET MODEL
7 CIP APPLICATION LAYER CIP PROFILE LIBRARY 6 CIP OBJECT LIBRARY CIP MESSAGE SERVICES 5 CIP CONNECTION MANAGEMENT	ETHERNET/IP	7 CIP APPLICATION LAYER CIP PROFILE LIBRARY 6 CIP OBJECT LIBRARY CIP MESSAGE SERVICES 5 CIP CONNECTION MANAGEMENT
4 UDP / TCP 3 INTERNET PROTOCOL (IP)		4 CAN MESSAGING 3
2 802.3 ETHERNET MAC		2 CAN LINK LAYER
1 802.3 ETHERNET		1 CAN PHYSICAL LAYER

Figure 40 - EtherNet/IP vs. DeviceNet Protocol Layers

CAN provides DeviceNet with several distinct application advantages:

- 100% bus utilization due to the bitwise arbitration discussed earlier
- Power on the bus means devices do not require any auxiliary power
- Simpler, more cost-effective components (but more expensive cabling than Ethernet)
- More adaptable for lower cost sensors and actuators
- Simplicity, ease-of-use and cost-effectiveness
- Miswiring protection
- Standard 24VDC Power
- Node insertion and removal under power

Is DeviceNet Obsolete or Complementary to EtherNet/IP?

DeviceNet, if not obsolete, is as least is on a downward trend. The open question is: will controller vendors continue to support it in the future or only provide access to protocols operating on standard Ethernet? If controller vendors introduce new controller lines that don't support DeviceNet, that would likely be the death knell for DeviceNet. In that scenario, device vendors, like motor drive vendors, would also discontinue support for DeviceNet (device vendors always follow the lead of the controller vendors). At that point, end users will have no choice but to switch from DeviceNet to an Ethernet implementation.

The other headwind that DeviceNet faces is component cost. As fewer devices are made and the technology ages, component costs increase. Ethernet is continuing to get less expensive as industrial Ethernet is no different to commercial Ethernet at the Physical Layer. If there is a tipping point where Ethernet devices become comparable or less expensive than DeviceNet devices, there will likely be an erosion of support for DeviceNet by end users.

Either way, manufacturing is not prone to radical changes. We should continue to have DeviceNet around to some extent for the next 20 to 30 years. It's over 20 years old now. Let's hope it sees a 40th or 50th birthday.

THE EVERYMAN'S GUIDE TO ETHERNET/IP

ETHERNET/IP AND RTA

Real Time Automation is an expert provider of consulting, services, training, and products for EtherNet/IP, DeviceNet and other CIP protocols. Our engineers have authored papers, given presentations and trained hundreds of engineers on CIP technologies. Our products and services can be classified into four categories:

DATA TRANSPORT

Often, there will be devices that must be integrated into an EtherNet/IP system that aren't EtherNet/IP enabled. When that happens, our broad line of EtherNet/IP gateway products can fit the bill. Our products have been successful for almost twenty years because of:

1. **Ease-of-use**: Our products are the only ones shipped pre-configured. That means if you order a Modbus TCP / EtherNet/IP gateway, it comes ready to run as a Modbus TCP / EtherNet/IP gateway. You get a product that never requires you to fight through a long series of configuration options that just don't apply to your application.
2. **In Stock**: When you hear "Always in Stock," you can count on it. Our products are manufactured in the US and are always available and shipped when you need them. Call prior to the end of the business day and we'll find a way to get it to you tomorrow.
3. **Outstanding Support**: Your phone call is answered by a live human and you'll be talking to a knowledgeable and empathetic support professional in less than a minute. And that's our promise! You can count on it.

DEVICE ENABLEMENT PRODUCTS

RTA was the very first company (yes, before Rockwell) to receive a Declaration of Conformance (DOC) for an EtherNet/IP Adapter (2002). Device vendors with products like drives, valves, linear actuators, barcode readers, visions systems and many other products use Real Time Automation software and hardware to enable their products for EtherNet/IP, DeviceNet and other CIP technologies.

ENGINEERED SYSTEMS

When you don't have the resources or the expertise to complete a product development, you can get assistance from our expert engineering staff. You can get us to do a little or a lot – it's up to you. RTA is your knowledgeable and trained team available to assist you with your project when you need it.

TRAINING

Twice a year you can attend training sessions on industrial technology. These two-day events provide in-depth discussions of EtherNet/IP, DeviceNet, PROFINET IO, EtherCAT and other factory floor technologies. They are perfect for a new Control Engineer to get acquainted with factory floor technology or an experienced engineer to learn the fine points – all in an unbiased, just-the-facts environment.

WWW.RTAUTOMATION.COM

ABOUT THE AUTHOR

John S. Rinaldi is Chief Strategist, Business Development Manager and CEO of Real Time Automation (RTA) in Pewaukee, WI.

After escaping from Marquette University with a degree in Electrical Engineering (graduating cum laude, no less), John worked in various jobs in the Automation Industry before fleeing back to the comfortable halls of academia. At the University of Connecticut, he once again talked his way into a degree, this time in Computer Science (MS CS).

John achieved marginal success as a Control Engineer, a Software Developer and IT Manager before founding Real Time Automation because "long term employment prospects are somewhat bleak for loose cannons."

With a strong desire to avoid work, responsibility and decision making, John had to build a great team at Real Time Automation. And he did. RTA now supplies network converters for industrial and building automation applications all over the world. With a focus on simplicity, US support, fast service, expert consulting, and tailoring for specific customer applications, RTA has become a leading supplier of gateways worldwide.

John freely admits that the success of RTA is solely attributable to the incredible staff that like working for an odd, quirky company with a single focus: "Create solutions so simple to use that the hardest part of their integration is opening the box."

John is a recognized expert in industrial networks and a speaker. He's spoken at events sponsored by the ODVA, Profinet International, the OPC UA Foundation and many industrial distributor events in the US and Europe. John is an author of seven books, hundreds of blog articles and scores of articles on communications and control in Automation magazine, Control Magazine and many others. John is a published article with five books on Industrial Networking.

John is famous for the articles in his Automation newsletter, one of the most widely read newsletters in the automation industry. The newsletter is a combination of John's opinions, technical insights, humor and fun. It's a must read for many automation professionals.

You can reach John here:

John S. Rinaldi
Real Time Automation
N26W23315 Paul Rd
Pewaukee, WI 53072
jrinaldi@rtautomation.com
262-436-9299 (Office)
414-460-6556 (Cell)

http://www.rtautomation.com/contact-us/
https://www.linkedin.com/in/johnsrinaldi

Have a little fun and get some relevant information. Sign up for the Real Time Automation Newsletter!

http://www.rtautomation.com/company/newsletter/

OTHER BOOKS BY JOHN S. RINALDI

OPC UA: THE BASICS

This book is a quick introduction to OPC UA for people who don't need to become experts but would like to talk knowledgably about OPC UA technology. It's not very detailed and in a few places mischaracterizes the technology, but it accomplishes its purpose as a basic introduction to the technology.

OPC UA: THE EVERYMAN'S GUIDE

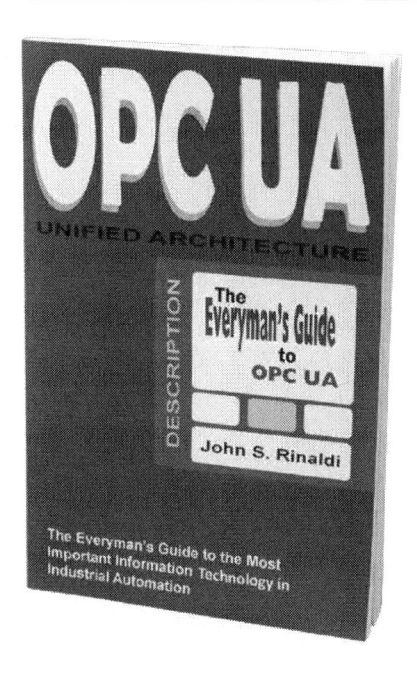

This is book for the dedicated professional that needs to really understand OPC UA technology. It's organized well and gives the reader what they need to know about OPC UA without losing them in the details. In fact, the book organizes the technical topics in such a way that the reader can choose to read just a quick overview of each topic, the concepts you need to know about that topic, or the full details.

MODBUS – THE EVERYMAN'S GUIDE TO MODBUS

In manufacturing automation, we use a lot of old technology. Yet even in our world, Modbus isn't just old technology. IT'S ANCIENT TECHNOLOGY. We live in a new age. The age of enterprise communications. It's an age where automation and the factory floor are changing in ways that weren't imaginable just a few short years ago. And despite all this, Modbus is still with us and is going to be with us for a long time. Modbus devices have permeated every kind of automation and will continue to over the next hundred years due to their simplicity and because Modbus is perfect for a lot of simple devices.

This book describes Modbus technology and the role that Modbus will continue to play in the future.

INDUSTRIAL ETHERNET

This book is an introduction to Industrial Ethernet. This book is considered the go-to guidebook for people who need to fully understand factory floor Ethernet and for those who need to have a basic understanding of Ethernet and TCP/IP terminology, Ethernet hardware, Ethernet software, Ethernet security, and the Internet of Things (IoT). The latest edition includes:

- The Industrial Internet of Things (IIoT)
- Ethernet topology
- Synchronizing devices over Ethernet
- Microsoft, Oracle, and Amazon Cloud platforms and web services
- The constraints of the industrial environment and the specialized requirements of machine control
- Power over Ethernet
- Wireless Ethernet
- Ethernet and IoT protocols and addressing

It also includes practical reference charts and installation, maintenance, troubleshooting, and security tips, which make this book an ideal quick reference resource at project meetings and on the job.

Made in the
USA
Columbia, SC